W9-DHX-863

ORDER NUMBER EA-352

AVIATION ELECTRONICS

By Keith W. Bose

International Standard Book Number 0-89100-352-5
For sale by: IAP, Inc., A Hawks Industries Company
Mail To: P.O. Box 10000, Casper, WY 82602-1000
Ship To: 7383 6WN Road, Casper, WY 82604-1835
(800) 443-9250 ❖ (307) 266-3838 ❖ FAX: (307) 472-5106
HBC0491 Printed in the USA

IAP, Inc.
7383 6WN Road, Casper, WY 82604-1835

Fourth Edition
First Printing — 1981
Second Printing — 1990

Library of Congress Catalog Card Number: 81-52156

Keith Bose is a technical publications consultant and writer specializing in electronics. He is the author of a wide range of articles, books, and manuals. His work has included airborne computers, guidance systems, telecommunications, and video systems. He served four years as an electronics instructor for the U.S. government, then specialized in aviation electronics for several years before devoting full time to technical writing and consulting. He is the author of a recent book on closed-circuit television.

Keith is a pilot and licensed ground instructor. He is a senior member of the Society for Technical Communications, and holds a bachelor of science degree from the State University of New York.

Contents

CHAPTER 1

CHAPTER 2

CHAPTER 3

CHAPTER 9

CHAPTER 10

CHAPTER 11

CHAPTER 12

CHAPTER 13

CHAPTER 14

CHAPTER 15

CHAPTER 16

Preface to
Third Edition

When the first edition of this book appeared, commercial airlines were in the process of phasing out propeller-driven aircraft for turbojets. Almost all aviation-electronic systems still used vacuum tubes. Now all aviation-electronic systems are produced with solid-state active components. Cockpit displays now use solid-state digital alphanumerics. Computer techniques have been adapted, and solid-state nonvolatile memory chips are making it possible to program cockpit computers for navigation and control.

General aviation has adapted to changing transportation requirements. Not so many years ago a light aircraft capable of no more than a 300-mile trip was still a valuable transportation tool. But the final development of the national interstate-highway system has made the automobile a strong rival for the 300-mile business trip, and light aircraft now must have better IFR capability and longer range for longer trips.

Electronic systems for general aviation now form a larger proportion of the total expenditure for an airplane. It is estimated that the average of all general aviation electronic system expenditures per airplane is near $17,000.

The first edition of this book was published as *Aviation Electronics Handbook*. The second edition was simply entitled *Aviation Electronics*. Over the years the term *avionics* has come to include all of the electronic systems used in aviation. This book covers the subject of avionics from the standpoint of the aircraft itself, rather than the ground apparatus that works with it.

A certain amount of theoretical detail with mathematical explanation has been included in this edition. This is for the benefit of readers who require a theoretical approach. In such cases there is an accompanying nonmathematical explanation for those who prefer this treatment.

Previous editions of this book have been adapted for teaching, reference, and general readership by pilots and aviation specialists. This new edition covers developments in the field resulting from new technology. It brings together information not readily available in a single source. Nevertheless, certain elements always remain. For example, the electromagnetics of Maxwell and Hertz are as unchanged as they were a century ago when first published. It is the new technology that makes new editions necessary.

Actual equipment has been chosen for illustrative description as examples of the art. These examples do not necessarily suggest the state of the art, nor optimum design. Rather, they are used to inform the reader of what to expect in actual equipment.

Thanks is again extended to all the firms who have contributed to this text since it was first published.

KEITH W. BOSE

Preface to Fourth Edition

The first edition of this text contained 11 chapters. The text now contains 16 chapters. This testifies to the growth of avionics technology and also to the requests of readers for more subject matter.

The wide acceptance of this text within the aviation community is acknowledged with this fourth edition. It has been expanded to include additional subjects and is again revised to keep pace with advances in electronic equipment that is now flying. Additional subject matter includes digital radar, newer avionics test equipment, and other material. The expanding use of RNAV has resulted in an additional chapter that also includes low-frequency global navigation.

Thanks is extended to the many firms who have cooperated in presenting state-of-the-art material. Most important was the gracious help of Mr. Charles May of Piper Aircraft Corporation who helped gather material.

KEITH W. BOSE

1

Standard Communication, Navigation, and Air Traffic Control Systems

The utility of the airplane is greatly limited unless all-weather operation is possible. Over the years, huge sums of money have been invested in the development and installation of electronic systems that make safe all-weather flight a reality.

The Common System of Air Navigation, Communications, and Traffic Control must be capable of satisfying the requirements of airline, corporation, private, and military flight operations. Growth of aviation all over the world is requiring constant advancement in techniques. At the same time, owners of expensive electronic equipment require some guarantee against obsolescence; hence, each development must be carefully "phased into" the common system. As the development of systems progresses, it becomes necessary to standardize characteristics for compatibility. Many government, quasi-government, and corporate organizations now exist to originate standards with the result that the entire Western world, with only a few exceptions, now operates under a common system.

The Federal government, through the Federal Aviation Administration (FAA), controls and maintains the ground facilities of the common system. This includes air traffic control (ATC) along with associated flight service. Maintenance of the vast facilities of the FAA is a task which goes on day and night. An example of on-board flight equipment required for operation is the flight deck of the DC-8 jet airliner shown in Fig. 1-1. Controls for the communication and navigational equipment are located on the throttle pedestal, between the two front seats.

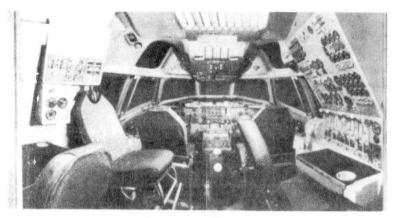

Fig. 1-1. The flight deck of a McDonnell Douglas DC-8 transoceanic jetliner.

This text covers the electronic equipment within the airplane. First, however, it is most logical to survey the FAA ground facilities which make up the common system, and then to examine the problems and responsibilities of the pilot.

RULES FOR FLIGHT OPERATIONS

When learning to fly an airplane, the student refers to the terrain below. As long as he is in visual contact with the earth, he is able to control and maneuver the airplane with his kinesthetic sense. When the pilot does not have an outside reference, his sense of control becomes distorted, and the airplane can go into a dangerous spiral. In the early days of flying, visual contact was the only means available. In the history-making nonstop flight of Alcock and Brown from St. Johns, Newfoundland, to Clifden, Ireland, in 1919, Brown told of the airplane descending violently while flying through the overcast of the Atlantic. Suddenly breaking out of the overcast, the two aerial pioneers found the airplane to be in a dangerous spiral. Fortunately there was enough altitude left for Alcock to recover, which he did skillfully.

Airplanes of today are equipped with gyro instruments to establish flight reference. Flying "on instruments" is more difficult than contact flying, and pilots are required to have an instrument rating before engaging in flight by reference to instruments. When visibility is good, flights are conducted under visual flight rules (VFR). During such flights, it is not necessary for pilots to use gyros for reference, and airplanes avoid each other on a "see-and-be-seen" basis. When visibility is below the minimum requirement for VFR, the flight must be conducted in accordance with instrument flight rules (IFR).

When operating VFR, the pilot is responsible for avoiding collision and can fly only when visibility permits. Under IFR flights in controlled air space and with poor visibility, the responsibility for avoiding a collision rests with the FAA air traffic control service. Prior to an IFR flight, a flight plan must be filed. Air traffic control then issues continuous instructions along the route until the airplane has safely landed. There are conditions where IFR and VFR flights can be mixed, in which case the IFR pilot is required to assume "see-and-be-seen" responsibility.

Existing and forecasted weather must be evaluated to determine whether VFR or IFR flight operations are required. Weather, therefore, is the most variable and complex factor in aerial operations. To gain maximum use of the airplane, new methods are constantly required to assist in all-weather flight operations.

There are three phases of an aircraft flight—takeoff, cruise to destination, and landing. When traveling from flight origin to destination, the pilot in command is obligated to choose the proper route, taking into consideration weather (including wind), fuel consumption, and aircraft performance. He must carry an emergency fuel supply above that calculated for the predicted flight. A pilot must know exactly where he is at any instant of the flight. During flight training, he first learns to know his location by reference to terrain (known as *pilotage*). Flying by reference to terrain was the only navigational means in the early days of aviation. Today the pilot identifies his position with on-board electronic equipment that will be discussed later in the book. He may also be located by ground devices, principally radar and radar transponders.

STRUCTURE OF THE AIRWAYS

When weather favors, a pilot may fly anywhere (except in certain areas) at anytime without clearance or without informing anyone of departure or arrival. He may, however, file a flight plan as a precautionary measure in the event of accident along the way, etc. When weather (visibility) requires, however, the pilot must fly IFR and the flight must be under the control of air traffic control centers. Also, any airplane flying 18,000 feet above sea level (MSL), must fly IFR and be completely equipped with a transponder and DME. This requires a systematic, standard means of conducting flights.

To make air traffic control possible, fixed air routes, known as *airways,* are laid out. Airways are channels in the sky defined by electromagnetic radiation. Ground equipment radiates signals in various forms that are interpreted by on-board electronic equipment and are used for the three phases of departure, en route, and terminal maneuvers.

The airways, as well as all other information necessary for IFR flight, are carried on approved charts. These charts are available from the United States government, and their use is explained in Advisory Circular No. 90-1A, entitled *Civil Use of U.S. Government Instrument Approach Procedure Charts.* Commercially produced charts are available from the Jeppesen Company, Denver, Colorado.

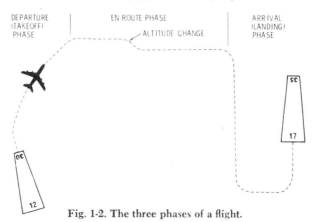

Fig. 1-2. The three phases of a flight.

Airways are numbered for identification. Points along airways are used for holding patterns, reporting points, etc. These points are given names that become very familiar to pilots, even though they are merely points in space marked by radio waves. Air route traffic control centers (ARTCC) are located throughout the nation, acting as nerve centers controlling IFR traffic. Each ARTCC has a specified area in which it assigns altitudes, receives in-flight reports, and otherwise exerts control. Each ARTCC is in contact with others. When an aircraft is within 30 minutes of the adjacent ARTCC area, an operator prepares to "hand over" the flight. Every effort is made to keep in radar contact with all IFR flights, and elaborate computer systems have been devised for traffic control.

Fig. 1-2 is a diagram of an IFR flight, showing its three phases. The departure phase consists of takeoff and climbing maneuvers. The en-route phase may be interrupted by an altitude change. Altitude changes are made by the pilot to get better winds, avoid turbulence, and so on. Air route traffic control requires altitude change to avoid other traffic or otherwise expedite traffic flow. The arrival phase consists of descent from cruising altitude, arrival maneuvers, including holds, and finally landing.

TERMINAL FACILITIES

Light aircraft often operate from small landing strips, where control of landing and takeoff traffic is unnecessary. At larger airports, a tower exerts visual control of traffic whenever possible, using radio to direct the pilots.

Tower Communications

For many years, a standard frequency of about 3023 kHz was used by low-powered aircraft transmitters to contact towers, with replies from the towers at a frequency of 278 kHz. Now, however, the vhf band is employed almost universally for such communications.

Towers are allotted frequencies in the vhf band from 118.00 MHz to 135.95 MHz. This band is broken down into small bands for specific purposes. For example, the frequencies between 122.00 MHz and 122.75 MHz are for communications from private aircraft to the tower. Towers answer on the band from 118.00 MHz to 121.40 MHz, as well as certain other frequencies. For ground-control purposes, towers use the band from 121.60 MHz to 121.90 MHz. A complete breakdown of the various vhf frequencies available, and their specified purposes, is given in Chapter 2.

If an airplane is suitably equipped, it may transmit on the same frequency as the tower. This is known as *single-channel simplex*. When each station transmits to the other on a different frequency, the system is called *double-channel* (sometimes, *cross-channel*) *simplex*. Single-channel simplex is used by towers on the ground-control frequencies to control aircraft taxiing from a landing or to a takeoff position.

Fig. 1-3 shows the interior of a typical modern control tower. Controllers utilize consoles from which the transmitters and receivers can be controlled. Equipment is fixed-tuned to operate on the assigned frequencies of the tower. Fig. 1-4 shows the exterior of such a tower. The receivers and transmitter are located in a space below the observation "cab." In some cases it is advantageous to locate the

Fig. 1-3. Inside view of a control tower "cab."

receivers and transmitters, together with their associated antennas, away from the tower. Such equipment is still controlled from the tower by special remoting equipment.

Precision-Approach Radar (PAR)

Precision-approach radar equipment is located near the runway and is especially designed to accurately locate an airplane within 300 feet of range, 10 feet of elevation (at a distance of one mile), and 20 feet laterally. The beam of a PAR system is swept alternately in the horizontal and vertical plane by switching back and forth

Fig. 1-4. Exterior view of the tower shown in Fig. 1-3.

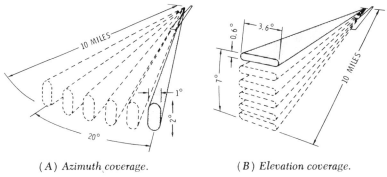

(A) *Azimuth coverage.*   (B) *Elevation coverage.*

Fig. 1-5. Beam coverage provided by a precision approach radar (PAR) system.

between a horizontal and vertical antenna. Beam coverage is shown in Fig. 1-5. In the standard PAR now used by the FAA, scan is provided without moving antenna reflectors. Because the sector of scan is fairly small, the waveguide antenna feed can be varied in such a way that the resulting beam is varied. The elevation and azimuth beams originate from reflectors which are 16 and 8½ feet long, respectively.

In the PAR indicator, both azimuth and elevation information are displayed as shown in Fig. 1-6. The elevation of the aircraft is determined from the upper portion of the display, and the azimuth position from the lower portion. When visibility is poor, skilled ground controllers use PAR to "talk" the pilot into a landing by directing the

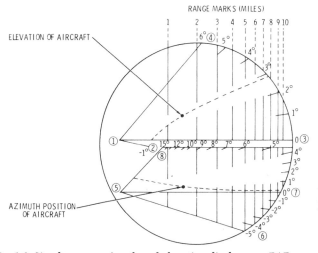

Fig. 1-6. Simultaneous azimuth and elevation display on a PAR scope.

19

aircraft within the limits marked on the indicator. The pilot, upon receiving instructions from the ground controllers, controls the airplane entirely by reference to instruments. Such a landing approach is known as ground-controlled approach (GCA).

Air-Surveillance Radar (ASR)

Air-surveillance radar allows controllers to "see" an area within a radius of 60 miles under any visibility conditions. Although lacking the higher resolving power of PAR, ASR has a greater range and gives a full 360° of coverage. Fig. 1-7 shows a typical air-surveillance radar installation. To a limited extent, ASR can also be used for ground-controlled approaches to a landing. However, the pilot must be able to break through the overcast at a higher altitude, since no control of altitude is possible and lateral control is less accurate than with PAR. Air-surveillance radar can usually bring an airplane to within one mile of touchdown.

All forms of precipitation—rain, hail, etc.—affect radar operation because some of the transmitted energy is reflected and appears on the radar screen, thus obscuring targets. Precipitation also absorbs and scatters a certain amount of the radar pulse energy, shortening

Fig. 1-7. A typical air-surveillance radar (ASR) installation.

the range. Precipitation effects are largely a function of frequency, and higher frequencies result in more pronounced attenuation. For example, radar operating at a wavelength of 20 centimeters provides about 12 dB less precipitation effect than one operating at 10 centimeters. Air-surveillance radar operates at 10 centimeters (3000 MHz), which is considered the frequency of optimum characteristics.

Man-made objects, and terrain features such as hills and mountains, return energy and appear brightly on a radar scope. When a radar is used for aerial surveillance, ground return clutters the display and obscures targets. Therefore, a system known as a *moving-target indicator* (MTI) has been devised which will display returns from only those objects in motion relative to the radar antenna. This system uses the important *Doppler* principle in which the frequency (hence, phase) of echoes returned from an object in relative motion will be shifted in phase from the returns from stationary objects. The system operates by comparing the phase of echo signals following one transmitted pulse with those following the next successive pulse, and canceling signals from stationary objects.

Fig. 1-8. Air-surveillance radar transmitting and receiving equipment located inside the building shown in Fig. 1-7.

Air-surveillance radar systems employ what is known as a *plan-position indicator* (PPI). The PPI scope presents a maplike view of the area surrounding the antenna. In the case of ASR, a system of video mapping using a photocell scans a map in such a way that a video signal is generated. This signal is then mixed with the radar video to superimpose a map of important boundaries, control areas, etc., on the regular video display. Fig. 1-8 is an example of the type of equipment found in the ASR building.

Airport Surface Detection Equipment (ASDE)

A third type of radar employed at airports is the airport surface detection equipment (ASDE). This radar system is especially designed for excellent definition and has a maximum range of 4 miles. It can detect airplanes separated by only 25 or 50 feet on the ground. Fig. 1-9 shows the appearance of an ASDE radar indicator. ASDE is especially useful at night, or on days of poor visibility, when towers cannot see to direct aircraft taxiing to and from the runways. Because the display condenses all ground traffic before a single observer, it can often be used to advantage over visual meth-

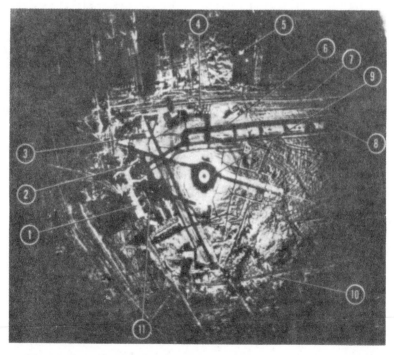

Fig. 1-9. Example of the pattern displayed on the PPI scope of an ASDE system. Notice how clearly the surface features are outlined.

ods. This equipment is only necessary at the larger, more congested airports.

VHF/UHF Direction Finders

Vhf direction finders were developed during World War II as an aid to fighter operations. Thereafter it was decided that the vhf direction finder could be used with good results in connection with ASR. Direction-finder antennas are usually of a type known as the *H-Adcock,* and are normally mounted at the ASR site. The direction finder automatically provides a reference signal from which the ground controller obtains a bearing to an airplane. The output is combined with the ASR scope presentation. A "strobe" or line of light points to the particular target transmitting on the chosen vhf/uhf channel. The direction finder is also used on the 121.5-MHz emergency channel to help locate lost aircraft.

The Instrument Landing System (ILS)

Landing in poor visibility by ground-controlled approach using PAR has already been discussed; however, the ILS provides another method of instrument letdown which is universally employed by commercial airlines. The instrument landing system consists of a radio beam, called a *localizer,* which is used to line the airplane up with the runway, and a *glide-slope* beam along which the airplane flies down to the runway. The glide-slope beam is inclined about 3°. Both glide-slope and localizer paths are erected by the use of two sharp lobes of rf energy which intersect at the correct path. One lobe is modulated at 90 Hz and the other at 150 Hz. The beam thus formed is received in the airplane, and the pilot uses a cross-pointer instrument which is centered when both 90- and 150-Hz signals are equal. Hence, the airplane flies down the path where the 90- and 150-Hz signals appear at equal strength.

Fig. 1-10 (see foldout at back of book) shows the complete ILS layout. An *outer marker* beacon is placed 4 to 7 miles from the end of the runway, and a *middle marker* is placed about 3500 feet from the runway, along the localizer path. The beacons transmit a 75-MHz signal directly upward. Each marker beacon has a different signal. The outer marker (OM) is modulated at 400 Hz and is identified with continuous dashes at the rate of 2 dashes per second. The signal from the outer marker automatically blinks a purple light in front of the pilot. The middle marker (MM) is modulated at 1300 Hz and is identified with alternate dots and dashes keyed at the rate of 95 dot/dash combinations per minute. The signal from the middle marker causes an amber light to blink. Thus, the pilot receives visual notice of his arrival over a definite point, and landing procedures are set up accordingly. Each airport will have an established *minimum ceiling* for each class of aircraft. If the runway cannot be seen above a certain minimum altitude as the airplane proceeds down the glide path, the pilot must abandon the approach, and proceed to another airport. In the critical last moments of a landing, after the airplane has broken clear of the overcast, special approach lights are erected to guide the pilot to the runway.

In addition to the outer and middle markers, some airports also employ an *inner marker* (IM) and a *back-course marker*. The inner marker, where installed, will indicate a point at which the aircraft is at a designated decision height on the glide path between the middle marker and the landing threshold. The inner marker is modulated at a frequency of 3000 Hz and is identified by continuous dots keyed at a rate of 6 dots per second.

The back-course marker, where installed, normally indicates the ILS back-course final approach fix where approach descent is started. The back-course marker is also modulated at 3000 Hz and is identified with 2 dots at a rate of 72 to 95 2-dot combinations per minute.

When using the ILS for terminal operations, landing conditions are broken down by Federal regulations into three categories:

Category I —200 ft ceiling (decision height), 2400 ft forward visibility.
Category II —100 ft ceiling (decision height), 1200 ft forward visibility.
Category III—zero ceiling, no forward visibility.

Category III landings are true "blind" landings. Although such landings have been accomplished for many years, equipment is not yet available that will allow a sufficient margin of safety for Category III commercial operations. Only a few airports are authorized for Category II landings. Airports with this category must have special ground-guidance lighting and "transmissometers." The latter is an optical device that determines forward visibility [visibility at the landing site is known as *runway visual range* (RVR)].

The Airport Site

Fig. 1-11 is a layout of a hypothetical airport capable of both Category I and Category II landing operations. In addition to the electronic equipment just described, approach lighting is required, along with lights to mark the runway, runway threshold, touchdown zone, and taxiways. Runways are laid out in the direction of prevailing winds. A runway is numbered according to its direction. For example, Runway 15 would have a direction of 150° from north. The other end of the runway (direction differs by 180°) would have a direction of 180° + 150°, or 330°, and would be known as Runway 33.

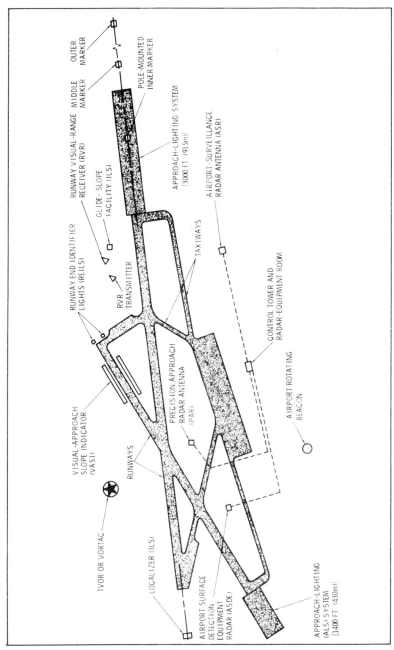

Fig. 1-11. Layout of a typical airport. Runways are constructed in the direction of prevailing winds for a given locality.

25

Landings under IFR conditions may be made without use of the ILS, but the ceilings required will be higher than 200 feet. These landings will vary for each airport, and for the landing method to be used. These are known as *nonprecision* landings. A nondirectional, low-frequency radio beacon may be used with the airplane's automatic direction finder. A low-powered VOR station located at the airport (TVOR) may also be used. In these cases, the pilot has only his altimeter to establish his descent, and the angular precision at which the runway is intercepted is much less than with ILS. Hence, the aircraft must break through the overcast at an altitude sufficient for the pilot to gain control by visual means and thus complete the landing.

EN-ROUTE FACILITIES

The discussion thus far has not included the facilities provided to guide the airplane safely along the route between the points of departure and arrival. There are two overlapping requirements for such flights: (1) The pilot must have a means of establishing the correct course and determining his location; (2) the FAA air traffic controller must control each flight to prevent collision. The latter is done by reserving a moving pocket of "airspace" for each airplane. On IFR flights, continuous communication with the airplane is mandatory.

Older En-Route Systems

The first en-route aids were the airway beacon lights established for the old Air Mail Service. Lighted federal airways were laid out connecting principal cities of the United States. Beacons were spaced at 10- to 15-mile intervals, depending on terrain, but were useful only when the ground was visible. They were replaced by radio guidance systems during the 1930s.

The low-frequency range system was developed by the Ford Motor Company. In 1927, two Ford-owned stations were located in Chicago and Dearborn. By 1928, the United States government began a program of installing 200- to 400-kHz radio ranges throughout the country, until a maximum of 400 had been installed by World War II. The stations were installed in such a manner as to connect the cities of the nation with aerial pathways. The low-frequency range system operates with an antenna array consisting of two loops at right angles and fed by a goniometer arrangement. A vertically polarized antenna, in the center of the array, emits a signal exactly 1020 Hz higher than the frequency of the signal radiated by the loops. The loops are keyed alternately with A and N ("dit-dah," "dah-dit") emissions. When exactly on course, the pilot hears a steady tone. If he hears a "dit-dah" or "dah-dit," he will be off course, but in a quadrant which can be determined by reference to his chart. There are four possible courses emanating from the station.

Lack of versatility, and the fact that low frequencies are subject to considerable atmospheric noise, caused the low-frequency range to be replaced by VOR (to be discussed shortly) as a means of en-route guidance. One thing in favor of the lf range, however, is that it can be utilized by airplanes equipped with only a simple receiver. Low-frequency stations no longer exist within the United States.

Nondirectional Beacons (NDB)

The nondirectional beacon is simply a 200- to 415-kHz transmitter operating into an omnidirectional vertically polarized antenna. Although one of the oldest radionavigational aids, it is still important. Airplanes equipped with automatic direction finders can tune to these stations, take a bearing, or "home" on them. Commercial broadcast stations can also be used for direction finding, but are difficult to identify because the pilot must wait for station announcements. NDBs continuously broadcast a 1020-Hz tone, giving a three-letter station identification in Morse code. Each NDB is identified and accurately located on the pilot's chart.

Nondirectional beacons are used to mark holding areas and obstructions. A low-powered NDB is usually placed at the ILS outer and middle markers, where it is known as a *compass locator*. Table 1-1 lists the five classes of nondirectional beacons.

Very-High-Frequency Omnirange (VOR)

Today, the very-high-frequency omnirange (VOR), sometimes referred to simply as *omni*, is the primary en-route guidance system in the United States. Very-high-frequency omnirange is not complex and has proven fairly reliable. Low-frequency range had only four courses to follow; VOR makes it possible to follow any course to or

Table 1-1. Nondirectional Beacons

Class	Power In Watts	Notes
HH	2000	
H	50–2000	
MH	25–50	
LOM	25	Outer marker; transmits two-letter identification.
LMM	25	Middle marker; transmits two-letter identification.

Fig. 1-12. A typical VOR station.

from a VOR station. Very-high-frequency omnirange transmitting frequencies range from 108.0 MHz to 117.9 MHz, which makes the system relatively free from the atmospheric noise encountered with lf range.

Very-high-frequency omnirange equipment is housed in a small circular building, as shown in Fig. 1-12. The location of each station is accurately indicated on the pilot's chart. To fly to or from any station within line-of-sight range, the pilot selects the heading relative to the VOR station. A deviation needle tells the pilot if he is to the left or right of the selected course or VOR heading. Most VOR stations are equipped with voice modulation. A pilot tuned to the station frequency will hear the identification signal in either voice or code, and also other important information. If he desires flight information, he may call the appropriate Flight Service Station° and the call will be answered via the VOR station. Thus, a single vhf receiver can serve for both communications and navigation.

Distance Measuring Equipment (DME)

The relative location of one point to another on a given level can be described by a minimum of two coordinates—direction and

---

°FSS frequencies are shown on charts.

28

distance. Very-high-frequency omnirange provides the direction, whereas the distance element is provided by DME. Together, VOR and DME provide all the information necessary for the pilot to fly the VOR airways. Both VOR and DME are housed in a single station called *VORTAC*. A pilot determines the distance to a station from the indication on a counter dial. Readings up to 195 nautical miles can be obtained. The DME was under development for several years, and began to go into full-scale commercial use in 1961. As more aircraft use DME, it will be possible to provide a smoother flow of traffic over the airways.

Air Route Surveillance Radar (ARSR)

Air route surveillance radar is the long-distance radar system employed by the air route traffic control center (ARTCC) to locate aircraft along the airways. Large airplanes may be detected up to 200 miles away. Plans are to cover the entire nation with a radar routing system. Radar equipment operating at remote sites provides video information, which is routed to the scopes at the ARTCC by microwave relay links.

FLYING THE AIRWAYS

The best way to get acquainted with the airways is to take a simulated flight. Suppose the flight is aboard a corporation-owned executive airplane. It is a Beech Model 88 Queen Air twin-engine executive airplane, registration number N7823B. Our departure point is Will Rogers World Airport, Oklahoma City; destination is The Greater Southwest International Airport, Dallas-Fort Worth, Texas. The airplane has been serviced and fueled.

The pilot has received a weather briefing and has determined that the flight must be IFR. Further briefing is received from the FAA Flight Service Station (FSS), and the pilot draws up his flight plan. He selects the direct route from Oklahoma City VOR to Bridgeport VOR (about 45 miles NW of Southwest International). This airway is marked on the charts as V-17. No later than 30 minutes before takeoff, the pilot files his flight plan with Fort Worth Air Traffic Control Center. The flight plan will give the type and registration number of the airplane, number of passengers, route, altitude, takeoff time, and other flight information. The entire flight will be under control of the Fort Worth Control Center, which controls the entire region. As soon as the flight plan is received by the Center, it is placed in a computer and a punch card is made up on the flight.

While passengers are boarding the airplane, the pilot and copilot are going through the preflight checklist. About seven minutes before scheduled takeoff, the engines are started and the tower is contacted on the ground-control frequency. The work in the tower "cab" is divided among *ground control* and *local control* operators. The ground-control operator will clear the flight for taxiing to take-off position, specifying the runway, velocity and direction of wind, and altimeter setting. The local-control operators have the task of exerting visual control over airborne traffic in the airport control zone, and of issuing instructions for VFR landings and takeoffs.

After ground clearance, the flight proceeds along the taxi strip to the run-up area near the end of the specified runway. There the engines are "revved" up and magnetos, generators, propellers, etc., are checked. The flight receives its clearance:

"ATC clears Beech 7823 Bravo to Bridgeport via V-17. Cross Alex at 13,000. Expect further clearance 10 minutes after take-off."

The point "Alex" is the intersection of the 318° radial from Ardmore, Oklahoma VOR with the 171° radial from Oklahoma City VOR. "Alex" intersection is therefore nothing more than a point in space marked by VOR intersections. In fact, the pilot's IFR chart does not show cities, terrain, and other geographic features, but only the various airways made up from the radio facilities appearing on the chart. (VFR charts do show terrain features, however.)

Having received clearance, the airplane is turned on to Runway 17 and the throttles are advanced for takeoff. As soon as the flight becomes airborne, it begins to appear on the ASR radar screen. The pilot contacts Will Rogers Radar Departure Control and is informed that radar has him in contact. In a few minutes the pilot reports that he has reached 5000 feet. The course is 210° magnetic. One VOR receiver is tuned to Oklahoma City VOR (OKC) and the other to Ardmore VOR (ARD). When the flight reaches airway V-17, which lies along the 171° radial from OKC VOR, Radar Departure Control is terminated. The flight is instructed to contact Fort Worth Center on a specified frequency.

The pilot calls the Center, giving his altitude, position, time, and destination; the Center replies:

"Beech 7823 Bravo cleared to Bridgeport. Maintain 13,000 feet."

The flight has reached altitude and will maintain cruising speed and altitude for 104 miles, or until Bridgeport VOR (BPR) is reached.

The crew has carefully set engine throttle and mixture controls for optimum cruise, and has "trimmed out" the controls. The autopilot is set to the "VOR Mode" with "Altitude Hold" engaged. The airplane now automatically flies along the VOR radial 171° "FROM"

OKC. If the airplane is equipped with DME, the distance from OKC will be constantly displayed.

As the flight proceeds, one cannot help but realize that it is totally dependent on the reliability of the ground facilities. The FAA makes periodic checks of all such facilities. Flight checks are made with airplanes carrying specially designed equipment. Maintenance programs are continuous, and ground facilities contain monitoring devices as an integral part. Should a VOR station fail, an automatic monitor alarms the ARTCC. NOtices To Air Men (NOTAMs) are sent out over teletype, so that pilots will not plan flights using these facilities.

About halfway between OKC and BPR, the pilot tunes to BPR and switches to the radial which will carry the flight along V-17. DME now reads distance to BPR. The frequency assigned by the control center must be constantly monitored. The Center calls:

"Beech 7823 Bravo, descend to 11,000, passing Nocona intersection at 11,000. Report leaving 12,000."

For some reason, ATC is moving the flight to a lower altitude. Possibly a military flight is crossing the airway, or other civilian traffic requires the airspace.

"Altitude Hold" on the autopilot is disengaged, power settings are reduced, and the descent begins. When the altimeter reads 12,000 feet, the pilot calls Center, states the time, and reports "Leaving 12." Nocona intersection is along the route, at its intersection with the 085° radial from Wichita Falls (SPS) VOR. The flight must be at 11,000 before reaching Nocona. Over Nocona, the pilot may report "Passing Nocona at 11."

BPR is a compulsory reporting point, and is so marked on the chart. As the flight passes over the BPR VOR station, the VOR needle swings violently, and the indicator swings from "TO" to "FROM." This places the airplane directly over the Bridgeport VOR station. Although only a small, isolated building on the Texas plains, this station represents the heart of the airways of the nation. The pilot reports "Over BPR." Center instructs:

"Beech 7823 Bravo cleared to Justin via Fox. Descend to 5000 passing Fox at 5. Report leaving 8."

Fox intersection is only 18 miles from BPR, and lies directly in line with the runway at Dallas-Fort Worth, 29 miles beyond. Justin is on the localizer, 18 miles from the runway.

The flight has been in darkness above a solid overcast. At cruising altitude the stars were brightly visible. In order to land, the flight must penetrate the overcast, maneuver to the outer marker of Greater

Southwest ILS, then begin a long approach along the glide path to the runway. As the airplane cruised on autopilot, the pilots were reviewing the procedure for the ILS letdown. The Center has directed an altitude which will be in solid overcast, and at 10,000 feet the airplane enters "the soup." Dallas-Fort Worth is reporting a ceiling of 200 feet and a visibility of one-half mile. This is just the minimum for an airplane of the Beech Model 88 class. The approach can be made, but if the approach lights cannot be seen when the altimeter indicates 768 feet mean sea level (200 feet above the runway), the landing must be abandoned.

Beech 7823B carries a radar beacon transponder. Because of this, the ASR at Dallas-Fort Worth has already positively identified the flight. Before the flight reaches Fox intersection, Center instructs:

"Beech 7823 Bravo, cancel your clearance to Fox. Contact Approach Control on 124.5."

Approach Control is located in the IFR Room, usually at the airport (see Fig. 1-13), and contains the scopes of the ASR. The ASR has three indicators—Radar Arrival, Radar Departure, and Flight Advisory. A skilled controller handles the function at each station. Approach Control guides flights to the ILS outer marker. Departure Control carries departing flights to the airways. Flight Advisory is primarily a VFR service which guides VFR flights on request, advises of traffic, and so forth. Beech 7823B now appears as a bright spot of light on the radar screen. Fixed marks on the screen indicate the correct path the flight must take to the outer marker, where the approach must begin. The airplane must be at 2000 feet when over the outer marker. As the flight reaches Fox, Approach Control instructs:

"Beech 7823B, turn to heading 129, descend to 2000."

The crew now completes the landing check list. The airplane is slowed to approach speed, the flaps are dropped and the landing gear is lowered. A purple light begins to blink in the cockpit—over the outer marker—and the approach begins. The airplane is controlled by reference to the ILS crosspointer. The cockpit instruments are illuminated by soft red lights. Outside, everything appears inky black.

The copilot calls off altimeter readings. Approach Control periodically advises the flight of its position along the glide path. If the airplane does not break through at an altitude of 768 feet mean sea level (just over the middle marker) the landing must be aborted, and the landing gear retracted, flaps raised, throttles advanced to climb power, and the airplane taken up to 2000 feet on the same heading.

But suddenly, an amber light begins to blink—the middle marker. Almost at the same time, the approach lights appear. Carefully, the pilot controls altitude and speed by visual contact with the approach lights. A line of green lights passes beneath; this is the runway threshold. The copilot applies full flaps, and the throttles are closed.

Fig. 1-13. A typical IFR room.

Gradually, the control yoke is brought back, the airplane loses flying speed, sinks, and the wheels touch the runway. The nose wheel is then allowed to contact the runway and brakes are carefully applied. Now the airplane slowly begins to taxi to the passenger ramp in a cold drizzle. At this point, Ground Control takes over and, with the aid of ASDE radar and blue taxiway marking lights, guides the airplane to the proper runway exit ramp.

COLLISION AVOIDANCE

Considering the vast amount of air traffic in the skies of America and Europe, the incidence of midair collision is statistically small. Nevertheless, morally even one incident is considered too many. Thus, there will be an outstanding quest for some kind of aid to collision avoidance. On a routine basis, in poor visibility under IFR, collision is avoided both through the assignment of flight paths by ground control personnel and through the normal vigilance of pilots. A collision occurs when these latter means fail. Closure speeds of aircraft are great; therefore, any practical apparatus to aid in collision avoidance must signal rapid, positive action.

When a flight is in progress from one point to another at constant altitude, the flight path is said to be *rectilinear;* that is, it follows a straight line. Aircraft that have reached cruising altitude and are under ATC direction will be following a rectilinear flight path. The geometry for a collision course of two airplanes in rectilinear flight is fairly simple to set up. With reference to Fig. 1-14A, note angles θ and θ'. It happens that whenever these two angles remain constant, the speed and direction of both aircraft is such that a collision is imminent, as shown. To avoid collision, it would be necessary to devise some means of actuating an alarm when this condition is present. A condition such as this, however, would only occur as a result of either instrument failure or gross human error. This is, therefore, not a typical collision condition. More typical collision circumstances are when aircraft are maneuvering and changing altitude, wherein the path through airspace is not a straight line, but *curvilinear,* as in Fig. 1-14B.

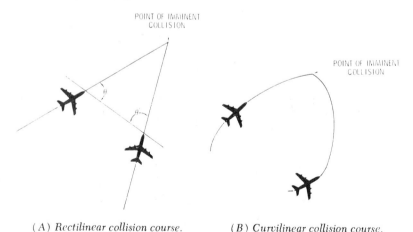

(A) *Rectilinear collision course.*   (B) *Curvilinear collision course.*

Fig. 1-14. Collision course of two airplanes.

We have pointed out that aircraft flight is divided into two classifications, visual flight rules (VFR) and instrument flight rules (IFR). When weather is such that visibility is above the prescribed minimums, an aircraft may still file IFR and be flown under ATC direction even though the pilot can operate without reference to instruments. Under these conditions, there will also be aircraft in VFR flight and there is no obligation that these latter flights be under ATC control. In this situation, both VFR and IFR pilots are required to avoid collision on a "see-and-be-seen" basis. It is possible, however, that many well-equipped high-performance aircraft can in fact be controlled with higher precision through reference to cockpit instrumentation than through reference to the horizon. Coupled to this, the IFR pilot usually may not bother to look out of the cockpit, but habitually may control by instruments through most of his flight at high altitudes where flights are exclusively IFR and there is less reason to observe ahead. At some point the IFR pilot may then enter a quasi condition where he will be obligated to visually monitor his flight path although still under an IFR flight plan. This condition occurs, unfortunately, at lower altitudes near airports where traffic is heaviest. Moreover, the cockpit team of the high-performance aircraft is then busiest with landing or takeoff procedures.

Here we have a situation where aircraft also are not necessarily in rectilinear flight, but may be maneuvering to include ascent and descent. This may create blind spots for one or both pilots, depending upon the configuration of the airplane. All of this describes the conditions where some sort of electronic aid would be of greatest value in collision avoidance. Yet this is the most difficult problem to solve by automatic means. The reason for this is that aircraft are in curvilinear flight paths which have an infinite number of variations that are difficult to predict.

This implies that the instantaneous location of aircraft relative to airspace must be known. Location by means of electromagnetic radiation suggests the most plausible means. In other words, the velocity and source of radio waves may in some way be employed for collision avoidance.

One approach to collision avoidance is the development of an infallible ground-based system using traditional radar principles. Systems now in use by the FAA have proven effective in many actual cases. There is a philosophical difference, however, between a ground-based system and a system that is self-contained within the aircraft. At first glance, an infallible self-contained system in every aircraft would seem most desirable. In practice, however, certain factors become important. It was pointed out that the geometric problem of collision avoidance is enormously complex. Systems within the present state of the art are very expensive even if effective. A simple ground system encompassing a given area would seem much more effective, and is in essence the end purpose of all traffic control in any case. Most important is the practical reality that the probability of collision is always greatest only during flight near terminal areas, where a single ground system is most feasible. Nevertheless, large funds have been committed in the past to airborne devices for collision avoidance.

Another problem of any self-contained on-board collision avoidance system (CAS) is the cockpit means of alerting the pilot, then conveying information for initiating the avoidance maneuver. It is not enough to warn of incipient collision, but the pilot must know whether to fly up, down, turn, etc. It must be noted that a professional pilot may fly for many years without arriving at a condition of imminent collision, possibly never. Yet in one brief moment, certain cockpit indicators must signal the emergency and dictate specific action of a highly unusual nature.

In studying the problem of collision avoidance, it becomes apparent that some means of evaluating closure between aircraft must be established. The distance between two aircraft divided by the rate of change of distance is equal to *time to collision* (or closest approach). This value is known as the *tau* (τ) *factor*, and is the major criterion for evaluating a collision threat. In collision avoidance systems under development contracts, one aircraft will compare its own altitude with that of all other cooperating aircraft. The systems command "climb/descend" if altitude and tau forms a threat or if altitude and distance alone signal a threat. If the CAS detects that the protected *altitude* boundary will be penetrated by another aircraft with either a tau or minimum range threat, the system commands "fly level."

The frequency band 1597.5–1622.5 MHz has been set aside for collision avoidance systems.

The basic equipment constituting the CAS includes an on-board digital computer, transmitting and receiving equipment, antennae to provide spherical coverage, and a timekeeping device. The heart of the system is the timekeeping device. Accurate timekeeping is essential because the system relies on measurement of the time difference from the start of transmission from one aircraft to reception by another aircraft, in order to determine the range between the two aircraft. Timekeeping accuracy to within one-quarter of a millionth of a second is obtained by atomic clocks at ground stations, and constant correction of less accurate airborne clocks by signals from more accurate clocks—either ground or airborne.

Transmissions from the various aircraft resemble a "count-down" —each number spoken by a different person. Only one aircraft transmits at a time, with all other aircraft listening. One complete sequence of aircraft reporting lasts three seconds and is known as an *epoch*. Each epoch is divided into 2000 time slots of 1500 microseconds in length. An aircraft transmits altitude and other data in a vacant time slot. The travel time of a transmission can be determined by the computer in the receiving aircraft once the time of reception is known, because the beginning time of each time slot is fixed by the system and is known by the computer. Thus, accuracy of timekeeping directly affects the accuracy with which the range of the intruding aircraft is determined.

The actual frequency used to transmit the short burst of digital information is kept precisely and is known also by the airborne computers. Hence, the apparent shift in frequency observed at the receiving aircraft is a measure of the range rate; that is, the rate at which the intruding aircraft is approaching. This is much the same as the *Doppler* shift in the pitch of a train whistle is a measure of the speed at which a train is approaching or departing.

This use of very accurate measurements of differences in precisely kept time and frequency is known as the *time-frequency* technique. This technique has made it possible to provide the required information with one-way transmission. By eliminating the need for two-way transmission, it is possible to accommodate the maximum number of participating aircraft with the minimum number of radio frequencies.

To determine a collision threat, the CAS computer determines the tau (τ) factor. Since new range and range-rate (or closing speed) information is received every 3 seconds, the computer makes a continuous succession of tau calculations. When tau reaches a predetermined value, the CAS commands the appropriate evasive maneuver. The time interval must be long enough to preclude the need for abrupt evasive maneuvers, yet not so long that it calls for too many maneuvers. The evasive maneuver required by the CAS is actually more gentle than an average jet takeoff.

For two aircraft in level cruise, the command to maneuver is given 30 seconds before a collision would occur. This warning time is the same for all types of aircraft; the difference is that the faster aircraft are farther apart when they receive the warning. Thus, two supersonic transports approaching head-on at a combined closing speed of 3600 mi/h would be 30 miles apart when the command to maneuver is given, while two piston-powered aircraft closing at 360 mi/h would be 3 miles apart when they receive the command to maneuver. For subsonic jets, the distance would be 10 miles, with a combined closing speed of 1200 mi/h.

The pilot of a CAS-equipped aircraft receives both visual and aural warning signals. Although different types of indicating devices have been proposed, all of them provide the following common commands:

Fly Up (Intruder is approaching, below your altitude.)
Fly Down (Intruder is approaching, above your altitude.)
Do Not Climb (There is an aircraft above you.)
Do Not Descend (There is an aircraft below you.)

Some CAS designs feature an alert warning to alert the pilot that he will soon receive a command to make an evasive maneuver. Another form of alert under consideration is one that tells the pilot not to climb or descend at a rate exceeding a specified number of feet per minute.

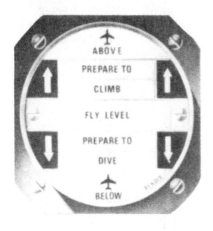

Fig. 1-15. The cockpit indicator of the Bendix collision avoidance system. (*Courtesy Bendix Corp., Avionics Division*)

The cockpit indicator of the Bendix CAS shown in Fig. 1-15 provides seven pilot instructions: (1) aircraft above, (2) prepare to dive, (3) dive, (4) aircraft below, (5) prepare to climb, (6) climb, and (7) fly level (in case the aircraft is climbing or descending when a collision threat occurs). Fig. 1-16 illustrates how the indicator functions to first alert the pilot and then finally give him a command to make an evasive maneuver. At the top of the illustration, the indicator "ABOVE" light is on, which tells the pilot that an aircraft is above him, but not necessarily in a hazardous position. In the center of the illustration, the indicator is warning the pilot that he is on a possible collison course with another aircraft and that he should "PREPARE TO DIVE." At the bottom of the illustration the "DIVE" light is on, which tells the pilot that he is on a definite collision course, with minimum safe time before collision of 30 seconds, and that he should immediately begin descending. The "DIVE" light will remain on until the threat of collision has disappeared.

Fig. 1-16. Illustration showing how the Bendix CAS indicator functions in one particular example to first alert the pilot and then give a command to make an evasive maneuver. (*Courtesy Bendix Corp., Avionics Division*)

In case an aircraft is climbing or descending and reaches a point of possible hazardous proximity to another aircraft, the indicator will advise the pilot to "FLY LEVEL" in order to maintain a safe distance from the intervening aircraft.

SUMMARY

The preceding is a discussion of what is known as the *common system* of air navigation and control, and the equipment used with it. The system is in worldwide use. It was developed and standardized first in the United States shortly after the end of World War II. Ground facilities of the common system are government operated, and the policy is that air facilities are to be used much as public highways. Within the United States and most countries there is no charge to an airplane that files a flight plan or otherwise uses ground facilities, although certain airports charge landing fees.

The common system has served remarkably well, and advances in technology are based upon those original concepts adopted over 30 years ago. Changes or additions to original concepts are approached carefully, sometimes among controversy, since this inevitably requires a large investment both by the government and aircraft users.

It is difficult to predict future developments in air navigation, because so much depends upon economic as well as technological factors. For example, the development of high-capacity jet transport has resulted in fewer commercial flights. Yet the multimillion dollar aircraft can economically afford a greater investment in electronics. On the other hand, general aviation, consisting of corporate and privately owned aircraft, is growing in number, and cannot afford the total investment in electronics that can be allowed for a commercial transport.

The foregoing factors are the criteria which govern changes in aerial operations. Other navigation methods have been used in the past and are in limited use for air navigation, particularly overseas. Low-frequency systems such as Decca and Loran are useful in over-ocean flights. Likewise, inertial navigation has been in use for several years in Boeing 747 aircraft. At present, however, such systems have little application within the continental United States, particularly where the investment per aircraft is relatively low in relation to that of commercial air carriers.

As a conclusion, it is apparent that the necessity to operate aircraft in low visibility weather, coupled with higher speeds, is the fundamental reason for the development of aviation electronics.

2

The Aviation Radio Spectrum

The radio-frequency spectrum has been divided into sections, or bands, within which a number of frequencies have been allocated for aviation purposes. Fig. 2-1 shows the bands used in the aviation service, and the uses for the frequencies within them. Some bands are better suited for a specific purpose than others. Before the reasons for this can be clearly described, however, a basic discussion of radio waves and their characteristics is appropriate.

RADIO WAVES AND THEIR EFFECT
AT AVIATION FREQUENCIES

Radio waves are the foundation of every radiocommunications and radionavigation system. By means of these waves, a pilot can determine his location, maintain an accurate course, and communicate with ground stations. Furthermore, the reflection characteristics of radio waves can be utilized to warn the pilot of storms and other weather conditions. These waves cannot be felt or touched—we recognize them only from their effects.

Radio waves are produced by a high-frequency alternating current that is fed to an antenna. The frequency of this current (produced by the radio transmitter) is determined by the number of times it changes its direction of flow each second. Generally, these

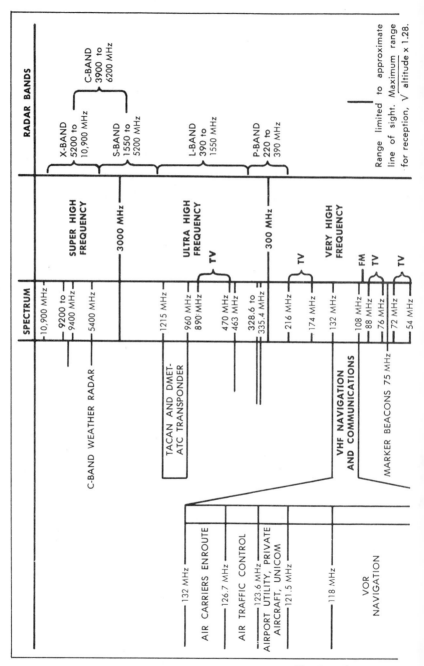

Fig. 2-1. The aviation

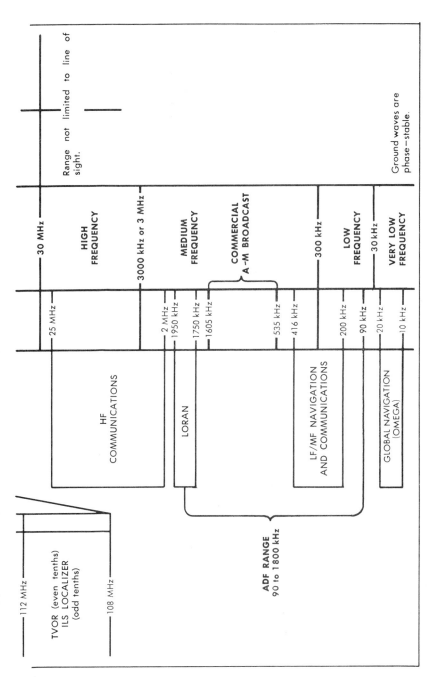

frequency spectrum.

frequencies are given in kilohertz and megahertz rather than in hertz. Thus:

1 kilohertz (kHz) = 1000 hertz.
1 megahertz (MHz) = 1000 kHz, or 1,000,000 hertz.

The energy in a radio wave is shared between an electric and a magnetic field, which are always at right angles to each other and continuously undergoing a change; the energy is alternately passed back and forth between these fields. As this occurs, field energy is traveling away from the antenna at the speed of light (186,000 miles, or 300,000,000 meters, per second).

The polarization of a radio wave is determined by the direction of the lines of force in the electric field, and the direction of these lines are in turn dependent upon the position of the radiating element. Hence, a vertical antenna emits a vertically polarized radio wave, and a horizontal antenna a horizontally polarized wave. In a receiving antenna, current is induced only by that component of the magnetic field at right angles to the antenna. Hence, under purely theoretical free-space conditions, a signal transmitted from a vertically polarized antenna could be effectively received only with an antenna of vertical polarization. Actually, this does not hold true under practical conditions, because the atmosphere and ground tend to break up the solid wavefront into components of varying polarizations. Practically speaking, the energy induced in a receiving antenna will be greatest when the sum polarization of the arriving wave is the same as the polarization of the receiving antenna.

Simple vertical antennas have omnidirectional characteristics. That is, they radiate and receive equally well in all directions. A horizontally polarized antenna, on the other hand, is basically directional (radiates and receives best in a given direction). Vertical antennas are generally used for radiocommunications, where operation is desired in all directions; whereas horizontal (directional) antennas find wide usage in radionavigation systems.

Also of importance are the propagation characteristics of radio waves. Waves may be reflected, refracted, or diffracted after leaving the antenna; their behavior is largely dependent upon the frequency and the type of transmitting antenna employed.

VERY-LOW FREQUENCIES

The characteristics of radio energy are greatly determined by frequency. At lower frequencies, the ground acts as a conductor. With ground acting as a conductor, a vertically polarized wave will induce currents into the ground that tend to "pull" the entire wave front along its surface. The wave front follows the curvature of the earth and is called a *ground wave.*

Very-low frequencies are only slightly above the audio range, and tend to follow the ground very closely. The wavefront moves around the curvature of the earth in a uniform front. The wave appears "phase-stable" as it travels through valleys and over mountainous terrain. This makes vlf valuable for certain types of radio navigation.

The phase of a wave at any point in space is determined by the relationship between the magnetic and electric fields at that point. Since the wave is in motion, there is a relationship between distance and phase. This is known as *wavelength,* and is

$$\lambda = \frac{c}{f}$$

where,

λ is the wavelength in distance units,
c is the velocity of radiation (units/second),
f is the frequency (hertz).

This is the distance traveled by a wave in one complete cycle. Thus, as the frequency decreases, wavelength increases. The wavelength is the distance from one point in a cycle to the same point in the next cycle. A very-low frequency of 16.188 kHz will have a distance of 10 nautical miles between cycles. A ground wave will therefore create distance markers around the earth. A means of measuring relative phase can therefore be used for navigation. This will be discussed in Chapter 5.

THE LOW AND MEDIUM AVIATION FREQUENCIES

The true effects of the ground on the propagation of radio waves are complex, principally because the ground appears different to waves at different frequencies. That is, the earth appears as a conductor at frequencies below 5 MHz, and as a dielectric above 5 MHz.

As the ground wave progresses, the resistive nature of the ground dissipates its energy. The amount of attenuation is dependent upon the nature of the terrain. As the frequencies increase, less energy follows the ground. A point is finally reached where ground-wave propagation is no longer significant. This can be seen by the graph in Fig. 2-2.

Frequencies below 1700 kHz are best suited for direction finding because there is minimum interference from reflections, and the ground-wave propagation is relatively good. Aeronautical four-course range stations and beacons operate on frequencies from 200

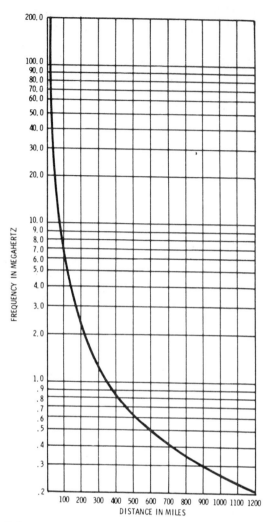

Fig. 2-2. Relative effect of signal frequency on ground-wave propagation for a given amount of radio-frequency power.

to 415 kHz. Marine beacons, which can be used by aircraft, operate from 285 to 325 kHz.

Marine communications occupy the band from 415 to 535 kHz; and 500 kHz is the international distress frequency, used principally by ships. Commercial broadcast stations from 535 to 1600 kHz can also be used by pilots for obtaining a bearing. The two advantages of broadcast stations are their high signal strengths and the fact that they are available almost anywhere.

However, since there are a great number of stations, they are sometimes difficult to identify, particularly since considerable time may elapse before a station announces its next call sign. Moreover, brodcast-station antenna arrays are often beamed toward heavily populated areas and thus cannot always be received over all areas. Only those commercial broadcast stations most desirable for navigation are noted on aeronautical navigation charts.

THE HF AVIATION BANDS

Between 100 and 250 miles above the earth, in the area only recently penetrated by rockets, satellites, and man-carrying capsules, great regions of ionized gases are gathered. Ultraviolet energy from the sun dislodges electrons from the orbits in the atoms of helium, nitrogen, hydrogen, and other gases, thus forming positively charged atoms known as *ions*. Gases of different masses seek different levels, forming ionic layers that surround the earth in the shape of a giant sphere known as the *ionosphere*. The individual layers shift constantly from night to day and season to season, and have been classified according to characteristics. The layers of major concern to radio-wave propagation are the D, E, F_1, and F_2 layers shown in Fig. 2-3.

The velocity of radio waves through an ionized layer is slightly less than through ordinary air or free space. This gives rise to the phenomenon of refraction or bending, and ultimately the return, or reflection, of radio waves. Under certain conditions, radio waves traveling into the ionosphere are reflected back to earth. As they

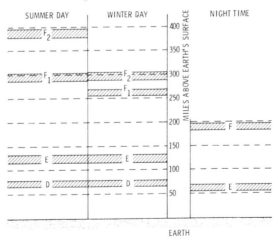

Fig. 2-3. Relationship between ionic layers of major concern during winter, summer, night, and day. The conditions shown in this illustration are typical for the Northern Hemisphere.

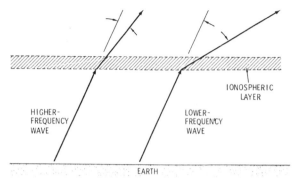

Fig. 2-4. Effect of radio wave frequency on the angle of refraction when the wave is passing through an ionized layer of the ionosphere.

reach the ground they are reflected upward again and back. The wave may be reflected to earth once or several times, depending on ionospheric conditions and frequency. In this manner a wave can be made to "skip" many thousands of miles, thereby making global communications possible.

Skip effect is explained as follows: When passing upward from one area into another where the propagation velocity is less, a wave will be refracted. The rules for this effect are described in physics by Snell's law. Waves entering the ionosphere are refracted at an angle depending on their frequency (see Fig. 2-4). As the angle at which a given wave enters the ionosphere becomes less, an angle is reached beyond which the wave is reflected to earth. This is known as the *critical angle* and is a function of frequency. Fig. 2-5 shows how radio waves at higher frequencies tend to be bent less, allowing all the energy to pass through the ionosphere and into space.

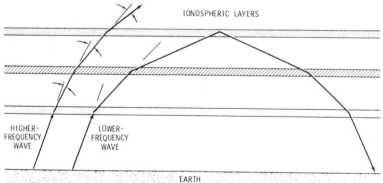

Fig. 2-5. Example of how higher-frequency radio waves pass through the ionic layers, whereas lower-frequency waves are reflected back to earth.

Fig. 2-6. A single-sideband hf transceiver. (*Courtesy General Aviation Electronics, Inc.*)

The height and nature of ionospheric layers can be predicted considerably in advance for each day of the year and time of day. With this information, it is possible to select a frequency which, in a single "skip" or series of "skips," will enable communication with any point—provided the proper antenna and sufficient power are used.

At one time, airplanes used only hf for communications. In the United States, however, hf is now employed in aviation for only long-distance communications by "skip." Until recently, trained radio operators responsible for tuning and operating hf equipment were carried aboard most long flights. Today, equipment is automatically tuned by the pilots. The desired frequencies are derived by switching combinations of crystals into the circuits, and servo loops tune the various resonant elements.

The Genave GSB/1000A (Fig. 2-6) is a small single-sideband hf transceiver. Communication on these bands is usually reserved for transoceanic flights. The longer wavelength requires either a trailing antenna or special loading. Also, the antenna must be tuned when a channel is changed.

The need for hf equipment in general aviation is not great, except in airplanes operating over water or in areas which lack vhf facilities. In smaller aircraft using hf equipment, a trailing wire antenna is sometimes employed. Its length is adjusted by a motor-driven reel to provide optimum antenna efficiency as frequencies are changed. In Mach 0.8 jets, however, external wire antennas cannot be tolerated. Instead, part of the airplane structure is arranged to form an antenna. Ingenious servo systems tune the antenna for optimum loading.

THE VHF AVIATION BANDS

As previously pointed out, radio waves of very-high frequencies normally are not reflected from the ionosphere. Also, ground-wave

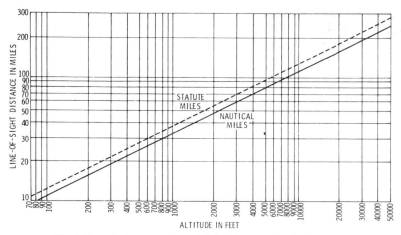

Fig. 2-7. Altitude versus line-of-sight distance for vhf reception.

propagation is relatively poor at frequencies within this range, as seen in Fig. 2-2. Theoretically, vhf radio waves afford "line-of-sight" communication only. However, these waves actually travel slightly beyond the horizon, because of the curvature of the earth's atmosphere.

For general air-to-ground work, the vhf frequencies are ideal. They afford positive communications with all points except those beyond a certain radius of interest, and thus minimize undue interference with distant, unrelated stations operating on the same frequency. An airplane at 30,000 feet has an absolute vhf range of approximately 200 miles. Very-high frequency has other good characteristics as an air-to-ground medium—much smaller antennas are required and power can be more effectively transferred from transmitter to receiver.

At an altitude of 1000 feet above a station, a maximum receiving distance of approximately 39 miles can be expected. Fig. 2-7 shows the vhf line-of-sight reception distances for various altitudes.

The designated aviation vhf band is from 108.0 to 135.95 MHz. Localizers and low-powered omnirange (navigation system) stations are alternately assigned frequencies between 108.1 and 111.9 MHz. The more powerful en-route vhf omnirange (VOR) stations operate from 112.0 to 117.9 MHz. The vhf communication frequencies run from 118.0 to 135.95 MHz, and are assigned in increments of 50 kHz (0.05 MHz). Frequencies from 118.0 to 121.4 MHz include towers, departure control, etc. The universal emergency frequency is 121.5 MHz and is isolated on either side by a "guard band." The 121.6- to 121.9-MHz frequencies are used for airport ground control. The FCC frequency assignments are tabulated as follows:

Air Navigation Aids

108.1–111.9 MHz: ILS localizer with or without simultaneous radiotelephone channel operating on odd-tenth decimal frequencies (108.1, 108.3, etc.).

108.2–111.8 MHz: VORs operating on even-tenth decimal frequencies (108.2, 108.4, etc.).

112.0–117.9 MHz: Airway Track Guidance (VORs).

Communications

118.0–121.4 MHz: Air Traffic Control Communications.

121.5 MHz: Emergency (World-Wide).

121.6–121.9 MHz: Airport Utility (Ground Control).

121.95 MHz: Flight Test.

122.0 MHz: Flight Service Stations, Weather, Selected Locations, Private Aircraft, and Air Carriers.

122.1 MHz: Private Aircraft to Flight Service Stations.

122.2, 122.3 MHz: FSSs, Private Aircraft, Selected Locations.

122.4, 122.5, 122.7 MHz: Private Aircraft to Towers.

122.6 MHz: FSSs, Private Aircraft.

122.8, 123.0, 122.85, 122.95 MHz: Aeronautical Advisory Stations (UNICOM).

122.9 MHz: Aeronautical Multicom Stations.

123.1 MHz: Search and Rescue (SAR) Scene of Action.

123.05 MHz: Aeronautical Advisory Stations (UNICOM), Heliports.

123.15–123.55 MHz: Flight Test.

123.3, 123.5 MHz: Flying School.

123.6 MHz: FSSs, Airport Advisory Service.

123.6–128.8 MHz: Air Traffic Control Communications.

128.85–132.0 MHz: Aeronautical En-Route Stations (Air Carrier).

132.05–135.95 MHz: Air Traffic Control Communications.

THE UHF BANDS

Beyond 300 MHz, drastic changes begin to take place in the methods of generating and radiating radio waves. Above 3000 MHz, transmission lines must be replaced by waveguides, and tuned circuits become resonant cavities. As the frequencies come closer to those of light, radio waves begin to take on the characteristics of light. The radar antennas used in aircraft are parabolic reflectors comparable to the reflectors of automobile headlights. Because uhf allows the use of parabolic reflectors or other means of focusing, it is possible to concentrate radio energy into narrow beams. This allows radar to transmit a burst of energy, and then to receive its return echo using the same parabolic reflector. The effects of refraction are more pronounced at uhf, and the atmosphere has a greater effect on propagation. The "index of refraction" of the atmosphere varies widely, doing so gradually and continuously throughout the atmosphere. Changes in humidity and temperature are the main reasons for variations in the index of refraction. In general, radio waves are turned back toward the earth. But occasionally the atmosphere forms "ducts," which act as a kind of waveguide and conduct the radio waves far beyond the horizon.

Military aircraft are the only users of the uhf spectrum for air-to-ground communication, employing frequencies up to 450 MHz. Glidepath-facilities, used with instrument landing systems, operate between 329.3 and 335 MHz. Surveillance radar and radar transponders operate from 1000 to 4000 MHz. Doppler-radar navigation systems, used aboard long-range jets, operate at 8800 MHz.

When meteorological conditions cause storm structures to form, the water droplets around the storm zone reflect energy quite well, making it possible to detect storms and turbulence with radar installed aboard aircraft. This type of weather-avoidance radar operates at frequencies up to 10,000 MHz.

AIRCRAFT ANTENNA REQUIREMENTS

Wire Antennas

Airplanes which cruise below about 300 miles per hour generally use wire antennas for hf, for "sense antennas" required with lf/mf automatic direction finders (ADF), and for 75-MHz marker-beacon reception. The characteristics of wire antennas will vary greatly, depending on their location, the size and shape of the airplane, and the operating frequency. The radiation pattern is composed of both a horizontal and a vertical component, and the magnitude of either varies independently with direction.

To minimize pickup of ice loads, wire antennas should be run parallel to the line of flight, or at not more than a 20° angle. Vibrational stress on airplane antennas is severe, and mechanical integrity must be maintained.

The trailing-wire antennas mentioned previously are suitable for small or slow aircraft, particularly where the hf radio is used only occasionally. Although this type of antenna is rarely used nowadays, it is worthy of mention.

Fixed-tuned hf antennas require tuning devices, to ensure maximum radiation at each frequency. When only a few channels are to be used, tapped circuitry representing variations in antenna length can be employed. Taps are selected by means of relays as the

Fig. 2-8. Vertical-type communications antenna mounted on the underside of a small airplane.

Fig. 2-9. Horizontal rod-type "vee" antenna mounted on the vertical tail fin of a small airplane.

transmitter is switched from one frequency to another. Most sophisticated installations use a servo-tuned device, which tunes the antenna to each frequency automatically.

VHF Antennas

At 100 MHz, the length of an efficient quarter-wave antenna is only about 30 inches. These dimensions permit such an antenna to

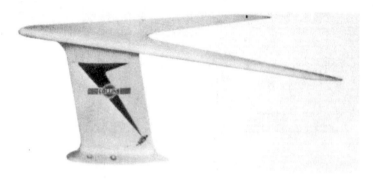

Fig. 2-10. A combination of two vhf antennas within a single structure. The lower portion forms a vertical antenna for communications purposes; the navigational "vee" antenna is in the sweptback horizontal structure. (*Courtesy Collins Div., Rockwell International*)

be mounted in a number of ways. Fig. 2-8 shows a vertical vhf antenna mounted on the underside of a small plane. This simple, inexpensive rod antenna gives good service on light aircraft, but obviously would be inadequate at higher speeds because of the greater mechanical stress.

Fig. 2-11. A structure containing a vhf communications antenna designed for high-speed aircraft. (*Courtesy Collins Div., Rockwell International*)

It is important that communications antennas be of the vertical types designed for omnidirectional service, to permit the reception and transmission of signals equally well in all directions. Vhf navigation signals generally are horizontally polarized, and navigation antennas must therefore be designed accordingly.

Fig. 2-9 shows a horizontal "vee" navigation antenna installed on the vertical tail member of a light aircraft. Such an antenna is electrically similar to the familiar television "rabbit ears." Unfortunately, this "vee" antenna cannot be made completely unidirectional. This is not a serious disadvantage, however, since the primary requirement is usually for reception along the longitudinal axis of the airplane—which it does provide.

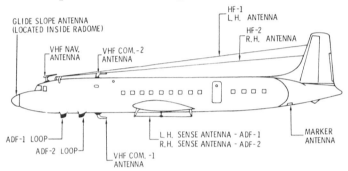

Fig. 2-12. This diagram shows the location and purpose of antennas used on airplanes designed for speeds up to about 300 miles per hour.

Rod antennas of this type are obviously suitable for low speeds only. The better antennas are composed of metallic elements enclosed in a highly streamlined fiber-glass structure designed to offer minimum drag. The elements are kept as small as possible, limited only by the physical size required to provide optimum operation over the intended range of frequencies. Figs. 2-10 and 2-11 show two such vhf antennas. The one in Fig. 2-10 is a combination of two vhf antennas built into a single structure. The vertical support forms the vertical antenna for communication purposes, and the navigation "vee" antenna is contained in the sweptback horizontal structure at the top. In Fig. 2-11 is shown a structure containing a vertical vhf communications antenna only. This type is commonly used on jet transports. While small aircraft may have only one or two antennas, larger planes generally employ many. The antenna complement of a commercial airliner is shown in Fig. 2-12.

3

Aircraft
Communication
Systems

Radio first entered the cockpit during World War I—the days of spark-gap transmitters and "brass-pounders" operating in the low-frequency range. These first radios were a crude beginning toward present-day lightweight equipment. Early air-to-ground communication was by means of cw (radiotelegraphy) only. By the early 1930s radiotelephone apparatus became reliable enough for air-to-ground use and was also much lighter than earlier equipment. Before World War II, practically all communication was within the lf and hf bands. Airplanes transmitted on the hf band, and ground stations answered on the lf band. For longer-distance communication, ground stations also transmitted on the hf band. At that time the standard answering frequency for towers was 278 kHz, and because the low-powered ground waves could be heard only in the vicinity of the airport, there was little interference. The design advantage of a single-frequency receiver is obvious. A disadvantage of low frequency, however, is high atmospheric noise level. When good, inexpensive vhf equipment became available, the low and medium frequencies fell into disuse.

During the 1930s research indicated that the vhf band offered the better means of air-ground and air-to-air communication, and the decision was jointly made to utilize the band of vhf frequencies now in use. The characteristics of vhf have been previously discussed in Chapter 2.

AVIATION COMMUNICATION FUNDAMENTALS

All forms of communication are accomplished by the transfer of energy between two points. This energy may be mechanical (i.e., sound) or electromagnetic (i.e., light or radio waves). Radio energy radiates outwardly from the antenna, constantly spreading out and occupying more and more space. As this happens, the amount of energy confined in one unit of space becomes less, as an inverse function of distance. This gives rise to the following rule:

The field energy varies inversely as the square of the distance from the source.

This rule tells us that the power in a radio wave drops off exponentially from the source. Another rule follows from the preceding:

At points at which the power is decreasing by a constant ratio, the distance is increasing in constant increments.

The attenuation of a wavefront is also affected by other sources, such as the resistivity of the ground (in the case of ground waves), atmospherics, and so on. Nevertheless, the aggregate result is always exponential and follows the rules given.

All through science, this proportionality of growth and attenuation persists. Ernst Heinrich Weber (1795–1878) of Leipzig University discovered the principle, now known as *Weber's law*, which states:

When stimuli increase by a constant ratio, the sensations aroused by them increase by equal increments or steps.

In other words, if an airplane is flying toward a station, and the pilot resets the gain of the receiver each time he notes an increase in signal, the points on the ground over which each resetting was made will be equidistant. This is the principle of the old lf range orientation procedure.

In order to simplify calculations, a unit known as the decibel (dB) has been adopted. The decibel is *not* a quantity of power or energy; it is a unit equivalent to ten times the logarithm of the *ratio* of two given quantities of power, or:

$$dB = 10 \log \frac{P_1}{P_2} \qquad (\text{Eq. 3-1})$$

A mere statement such as "10 decibels of power" is meaningless, since a ratio does not have dimensions.

One advantage of the decibel is that it can be used to evaluate a communication path. Mathematically, attenuation is expressed by a negative sign and gain by a positive sign. Thus, an amplifier may have positive gain, but the radio wave path through space will have negative gain. If a transmitter with a 1-watt output sends a radio wave along a 20-mile path at an attenuation of 2 dB per mile, the signal will be attenuated −40 dB. However, if a receiver with a gain of +40 dB is placed at the end of the 20-mile signal path, it will have an output of 1 watt. Therefore, if the algebraic sum of decibels along a path is zero, the circuit will have no loss. 'As a result, the output will equal the input.

In more specific terminology, the decibel is an accepted measure of the *response* to power or energy in terms of logarithms referred to our standard decimal number system. Hence, by this definition the term:

$$dB = 10 \log_{10} \frac{P_1}{P_2} \qquad (Eq. 3-2)$$

can be expanded by stating the equivalents of power in terms of voltage, current, or resistance, and we can write:

$$dB = 10 \log \frac{I_1^2 R_2}{I_2^2 R_1} \quad \text{or} \quad 10 \log \frac{E_1^2 R_2}{E_2^2 R_1} \qquad (Eq. 3-3)$$

Of course any expression equivalent to power would be permissible. We can simplify the previous expression for special cases. For example, suppose we have a device wherein the input and output impedances are equal, and we wish to refer input to output as an expression of gain:

$$dB = 20 \log \frac{I_{out}}{I_{in}} \quad \text{or} \quad 20 \log \frac{E_{out}}{E_{in}} \qquad (Eq. 3-4)$$

This expression is found often in specification work, and it is important to note that it is valid only when resistances in the terms for power are equal. Under this assumption, when relative power is doubled, we have the following:

$$dB = 10 \log 2$$
$$= 3 \, dB \qquad (Eq. 3-5)$$

This is because the logarithm of 2 is 0.3 ($10^{0.3}$). From this we note that the "half power point" is separated from some full value for power by 3 dB. When we speak of voltage or current in Eq. 3-2, a "half voltage point" or "half current point" would have a 6-dB separation (2×0.3). In circuits which casually or deliberately are frequency sensitive, the response can be expressed by stating the frequency band between the half voltage or half power points, as in Fig. 3-1.

In order to communicate over a greater distance, either the power of the transmitter or the gain of the receiver must be increased. Obviously, there is an economical and practical limit to the size of the transmitter. Receivers can be designed with amplification factors in the millions, but such large amplification means that tiny random circuit noises will also be amplified with the signal, producing a

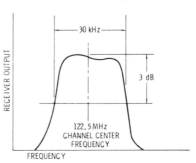

Fig. 3-1. Selectivity curve of aircraft receiver tuned for 122.5 MHz. Voice sidebands occupy 6 kHz, with an additional 24 kHz to allow for frequency drift.

roaring or rushing sound from the speaker. If the level of a minute signal received from a distant station is less than the noise voltage, increasing the gain of the receiver will be of no benefit. Thus, the gain of any receiver is limited by the amount of random noise present at all points up to and including the input of the first amplifying stage.

The origin of random circuit noise is attributed to molecular and atomic activity that is present at all times and has been found to be a direct function of temperature. The higher the temperature of any physical element, the greater will be the random noise present. Such noise has been found to possess infinite bandwidth. It is possible to calculate noise based upon theoretical value and this value is borne out quite closely in measurement. Noise originates both in the antenna and in the first input elements of the receiver. An expression for noise under these circumstances would be

$$N = F\kappa T \qquad\qquad (\text{Eq. 3-6})$$

where,

N is the noise power in watts,

F is a noise factor dependent upon characteristics of the individual receiver,

κ is Boltzmann's Constant (1.38×10^{-23}J/K),

T is the absolute temperature (Kelvin).

If we consider operation around room temperature, most receivers will be found to have noise factors of 1 to 20. The noise factor is of interest in design; however, for practical performance measurement, the sensitivity of a receiver is specified as signal-plus-noise to noise, expressed in decibels.

Receiver sensitivity is basic to the ability to receive a distant transmitter signal; however, there are other factors for any given installation that reduce performance. We may call this *receiver environmental loss*. Ambient electromagnetic noise at the airplane, antenna characteristics, and losses in connectors and transmission lines combine with intrinsic receiver noise to set a lower performance figure. The difference between the receiver's inherent sensitivity and the signal level required for tolerable performance is the measure of a given installation.

A government-operated ground station can be presumed to have a signal strength which at all points within the service area is adequate for the purpose. The signal will not uniformly attenuate from the station, however, due to terrain effects, altitude, and other causes. The relationship between receiving antenna and polarization characteristics and direction has been pointed out in a previous chapter. Losses in cables, etc., are apparent. Less apparent, however, is specific ambient noise. Noise originates within the engine compartment, power converters, and so on.

Noise is energy that has no defined frequency. That is, it is present at all points of the electromagnetic spectrum. Noises resulting from the previous analysis (Eq. 3-6) can be said to have infinite spectrum. In individual situations, however, noise can be greater at certain frequencies. Ignition sparks tend toward the very-high frequencies, although lightning in the atmosphere creates noise in the low-frequency range. In any case, noise can be reduced in a circuit simply by making the circuit sensitive only to the desired frequency.

SELECTIVITY

When communication is carried on through the amplitude modulation of a given carrier, the receiver is adjusted to receive only the band of frequencies occupied by the carrier and its sidebands. All other parts of the spectrum are rejected. This is the process known as *tuning*. The degree of rejection is a figure of merit known as *selectivity*. The superheterodyne principle, long used in household receivers, affords the best means yet devised for obtaining selectivity. In vhf aircraft receivers, double and sometimes triple conversion is used to obtain good selectivity.

Selectivity appears graphically as a curve in which frequency is shown as the independent variable with signal output of the last detector plotted on the vertical axis. The curve will demonstrate the ability of a receiver to pass desired signals and reject all others, as well as noise energy appearing outside of the passband. Fig. 3-1 contains a typical selectivity curve. The bandwidth of any frequency selective electronic device is considered to be measured between the two points on either side of maximum at which power falls off to one-half of the maximum. These points are known as *half power points,* and are therefore down 3 dB from the maximum.

It is in the if amplifier of an aircraft receiver that almost all of the gain and selectivity is provided. Gains of at least 40 dB must be obtained. The selectivity values shown in Fig. 3-1 are based upon the characteristics of aircraft communication signals. Intelligible voice signals range up to 3 kHz. When a carrier is amplitude modulated with 3 kHz it will occupy a 6-kHz band because modulation produces a 3-kHz upper sideband and a 3-kHz lower sideband. The center frequency is the frequency of the carrier. The carrier frequency of an aircraft transmitter is allowed to drift a maximum of 0.005% on either side of the established channel frequency. Also, the crystal stabilized local oscillators of the receiver may be assumed to drift by 0.005%. Thus, the bandwidth of a receiver must allow for carrier sidebands, a carrier drift, and local oscillator drift, resulting in a 30-kHz channel allowance. The receiver should be aligned so that its selectivity curve coincides exactly with the prescribed channel.

COCKPIT AUDIO REQUIREMENTS

Voice communication requires that attention must be given to audio reproduction within the cockpit environment. The pilot's microphone system must reproduce a voice with clarity while at the same time rejecting unwanted sounds. The pilot will choose either to wear earphones or to depend upon a cockpit speaker. More elaborate systems must include a means of switching the various audio circuits and possibly a cabin intercom or ramp hailer.

Cabin speakers of the automotive type serve well in aircraft environment and are available at reasonable cost. Microphones, however, vary widely in performance and deserve attention when procuring an installation. The microphone transmitting switch and certain other audio controls are used as often as the throttle, and placement of audio and communication controls requires thought and ingenuity.

Cockpit noise varies widely among different aircraft designs. Noise originates from the following sources:

Engine and propeller.
Vibration.
Air passage.

This audible noise will have components ranging from as low as 30 Hz to the highest audio range. Turbochargers and jet engines have very-high-frequency components, and so does the rapid passage of air around fuselage and wings. Pilots' and controllers' voices can be clearly reproduced within a band of from 200 to 3000 Hz, and it may be desirable to filter out the spectrum on either side of this. Some microphones only reproduce this narrow audio band.

Sound itself, whether in the form of noise or intelligible impulse, is a longitudinal wave consisting of compressions and rarefactions of the air. Psychologically, sound is sensed when these forces cause the eardrum to react. The pressure of sound, or its intensity, can be measured by units of force per units of area. Dynes per square centimeter is the usual unit of measurement.° It must be pointed out that this unit of pressure when so expressed does not specify frequency. There may be any combination of frequency elements included. Sound levels within the cockpit typically may run as high as +90 dB above 1 dyne/cm². Sound insulated cabins improve sound levels, particularly in single engine aircraft in which the fuselage mounted engine contributes to higher levels of sound. In all cases, the character of the sound, based upon frequency, defines the problem.

MICROPHONES AND HEADSETS

The oldest form of microphone still in use is the carbon type (Fig. 3-2). Carbon granules are arranged against a diaphragm in such a way that voice pressure causes diaphragm motion, in turn causing variable resistance to a direct current that is passed through the carbon granules. This type of microphone is a variable resistance device, as shown simplified in Fig. 3-3. It is characterized as possessing low impedance (the carbon granules average about 150 ohms), and fairly high output, about 2.5 volts rms. Resistance varies from 50 to 300 ohms when the diaphragm vibrates, depending upon sound level.

Current is fed in series with the carbon package in such a way that current remains fairly constant in spite of resistance variation itself. This means that the voltage drop across the carbon will remain a fairly linear function of diaphragm movement. It is accomplished by including a relatively large resistor in series as shown in Fig. 3-3.

Standard means of connecting microphones into the transmitter or audio system have been adopted. Fig. 3-4 shows the standard plug dimensions together with hookup. Note that aircraft microphones

---

°Under the new International System of Units, commonly known as the SI metric system, dynes per square centimeter (dynes/cm²) as a unit of sound pressure is replaced by the pascal (Pa); i.e., 1 dyne/cm² = 0.1 Pa.

Fig. 3-2. This Telex hand-held noise-canceling carbon microphone is a modern version of the type that has given long reliable service. (*Courtesy Telex Communications, Inc.*)

contain the push-to-talk key that activates the transmitter; however, it may also be placed on the aircraft control yoke.

All aircraft microphones and those intended for mobile communications use a noise-canceling principle. This feature is shown in the diagram of Fig. 3-5. Ambient cockpit noise is of a pattern that may be called *nonpolarized*. That is, it appears to originate from all directions. Of course, it originates from sources previously mentioned, but is echoed back and forth and fills the surrounding air. The pilot's voice, however, originates from a single point. By designing a diaphragm that responds only to a nearby point source of sound, unwanted sound may be canceled. For this reason, aircraft microphones must be held close to the mouth, often with the side of the mouth touching the instrument.

Another type of microphone that has come into use is the dynamic type. This type has a coil that is included within a magnetic field. Movement of the coil imparted by diaphragm motion generates a

Fig. 3-3. Principle of the carbon microphone.

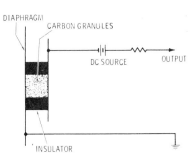

small current that is in direct relation to the modulating voice. Since the current generated is very small, a miniature transistorized preamplifier is included within the microphone. The amplifier can be made to operate on the same current used to energize a carbon microphone, and some are designed to be plugged into a transmitter built for carbon microphone input. By itself, the dynamic microphone has a much higher impedance than a carbon microphone.

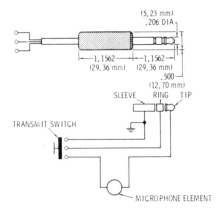

Fig. 3-4. Standard microphone jack.

Crystal microphones are another type and are now produced in large quantities. They have reasonably good qualities and are in wide use in amateur and Citizens band applications. This type of microphone uses two or more chips cut from Rochelle salt crystals. The chips are joined so that mechanical motion of a diaphragm causes a piezoelectric effect, generating a small current. A crystal microphone has a high impedance and will not operate unless plugged into a high-impedance input.

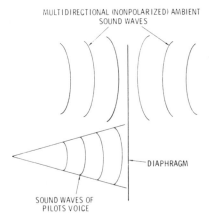

Fig. 3-5. Noise-canceling principle.

Fig. 3-6. An electret-type noise-canceling hand-held microphone.
(*Courtesy Telex Communications, Inc.*)

Another type of microphone is a descendant of the older so-called condenser type, now called *electret* type (Fig. 3-6). This microphone uses the change in capacitance between two plates caused by diaphragm action. The change in capacitance is relatively small; however, it faithfully follows the sound. The electret is a precharged dielectric that causes the capacitor element to appear charged. A preamplifier is also used with this type of microphone.

Headsets are generally of two physical types. Pilots may wear earphones of the type sold with popular transistor radios, or the

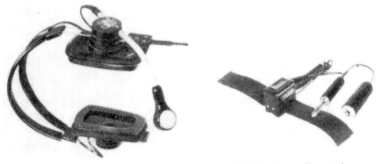

(*A*) *Microphone mounted on headset.*   (*B*) *Push-to-talk switch.*

Fig. 3-7. A combination dual-muff magnetic headset with mounted carbon microphone and push-to-talk switch. (*Courtesy Telex Communications, Inc.*)

larger headsets. Older models of headsets have a 2000-ohm input, however, newer models have been standardized at a nominal 600 ohms. In either case, headsets may be operated from speaker outlets but not vice versa. The most desirable arrangement, especially for IFR flights where a pilot must be in constant immediate touch with the controller, is to arrange the microphone on a boom together with headsets. The push-to-talk switch is then mounted on the control yoke (Fig. 3-7). A pilot may then use his radio without tying up one hand at crucial moments.

RECEIVERS AND CONTROLS

Receivers designed for airlines are usually mounted on the "electronic rack" and remotely controlled by the pilot. A standard vhf receiver control unit like the one in Fig. 3-8 is used for this purpose. In general, the frequency-selecting elements of both the transmitter and receiver consist of two crystal selector switches—one for 1-MHz increments and one for 100- or 50-kHz increments. The shafts of the variable tuning capacitors or inductors rotate with the crystal selector switches. A small electric motor drives the frequency-selector shaft to a position determined by the setting of the switches in the control unit. The motor drives the shaft through a positioning unit which accurately positions the shafts as selected by the control unit. Remote control is accomplished by a standardized binary open-seeking circuit known as the *reentrant system*. The total number of combinations of switch positions in such a system is 2N − 1, where N is the number of interconnecting wires between the control unit and receiver.

Lightweight receivers are mounted directly on the instrument panel, and switching is direct. In order to provide a lighter, smaller unit for the panel, sometimes only the rf portion of the receiver is

Fig. 3-8. Remote-control unit of a standard vhf receiver. (*Courtesy Collins Div., Rockwell International*)

mounted on the panel; the if, audio, and power supply are housed elsewhere. Although many receivers are capable of receiving both communications and navigational signals, a receiver is usually assigned only one primary function in a given airplane. Fig. 3-9 shows the panel view of a King KR-40 receiver. This unit is designed solely for navigation, and tunes from 108 to 117.9 MHz in 100-kHz increments. The front measures $2\frac{3}{8}$ by $6\frac{1}{2}$ inches, with a depth of $8\frac{3}{8}$ inches. The unit weighs less than 4 pounds, complete with power supply.

Fig. 3-9. Panel view of the King KR-40 navigation receiver.
(*Courtesy King Radio Corp.*)

SQUELCH OPERATION

An aircraft receiver always includes automatic gain control (agc), the familiar circuit universally used on household radios, television, and so on. A receiver equipped with agc will be operating in its most sensitive condition when not receiving a carrier. Under this most sensitive condition, the noise given by Eq. 3-6 will be greatly amplified and will be present in the speaker or phones. This is very annoying and fatiguing. If audio volume were to be turned down, a calling station may be missed. To eliminate this problem, a squelch circuit is employed to disable the audio circuit when no signal is present. The audio remains disabled until a carrier of sufficient level is received.

In the simple squelch circuit of Fig. 3-10, the minus agc voltage is used to control squelch. Assume that the higher the received carrier voltage, the more negative agc voltage becomes. Hence, in the absence of a carrier, agc voltage becomes positive, the base of npn Q1 goes positive, and the collector voltage drops across R2 and R3. At some point, the base of audio amplifier Q2 goes below about 0.6 volt, and the stage is cut off. This disables the audio amplifier and the cockpit speaker or phones are muted. Potentiometer R1 can be set so that Q1 reacts to a certain agc level. Thus, audio may be held in cutoff just at that point where noise disappears. When a carrier is received, agc voltage responds to the carrier voltage, and audio stages function. The squelch circuit described here is a simple example for explanation purposes. Several squelch methods are in use.

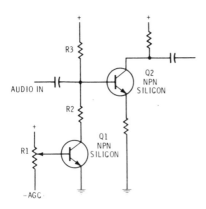

Fig. 3-10. A simple squelch circuit.

FREQUENCY SYNTHESIS

A continuously tuned receiver is one in which the local oscillator can be tuned continuously throughout a band by means of variable inductors or capacitors. Ordinary household receivers are of this type. The continuous-tuning principle is no longer used in aircraft receivers, however. The primary reason for this is that pilots would no longer be able to manipulate a tuning dial under conditions of modern flight. Receivers must be quickly tuned to an assigned frequency and changed often in flight, sometimes during very busy periods.

Frequency stability of aviation transmitters must be within ±0.005% of assigned frequency. Moreover, the mixer or local oscillator frequency must be capable of staying within the tuned intermediate (if) heterodyne frequency. Quartz crystal oscillators are the most economical means of obtaining the required frequency stability. To change frequency, crystals and other circuitry must be switched. Receiving and transmitting equipment must be capable of operating from 118.0 to 135.95 MHz in 50-kHz increments, or at least 180 distinct channels. Rather than employ 180 separate crystals, a means of obtaining stable frequencies from fewer crystals has been devised. This is known as *frequency synthesis.*

Two methods of frequency synthesis are now in use, the *heterodyne* method and the *phase-locked-loop* method. The heterodyne method is rapidly being replaced, since integrated circuitry makes the phase-locked loop much simpler, lighter, and more economical.

In both systems, the crystal-controlled oscillator is fundamental, but the phase-locked loop requires only one or two stable frequency sources which can be supplied by one or two crystals permanently

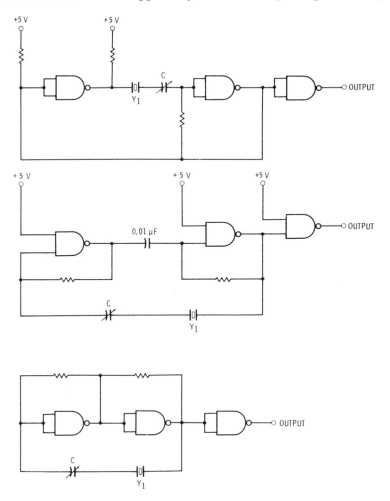

Fig. 3-11. Three forms of integrated-circuit crystal oscillators.

fixed within oscillator circuits. Almost an entire oscillator circuit may be contained on a single chip, with other chips forming the remainder of the phase-locked loop circuitry. Three forms of oscillator circuits are shown in Fig. 3-11. Note that each contains a variable capacitor. This is used to make minor adjustments to assure that the oscillator is at the correct frequency.

Heterodyne Frequency Synthesis

When alternating currents at two frequencies are present in a nonlinear circuit (a *nonlinear* circuit does not respond to Ohm's law), other frequency components are caused to appear. In essence, there will be a *sum* and a *difference* frequency. For example, if 100 MHz and 10 MHz are mixed, there will also be components of 90 MHz and 110 MHz present. Either of these may be filtered out and used as a distinct component. This is the principle used in heterodyne frequency synthesis. In this process, mixers are used to obtain the sum or difference frequencies of a combination of two or more oscillators in order to arrive at many frequencies, usually with digital relationship. For example, if it is desired to cover a band from 100 to 120 MHz in steps of 1 MHz, 20 crystals—one for each 1-MHz point from 100 to 120 MHz—are required. Nine other crystals for 100 to 900 kHz can be used to provide coverage in 0.1-MHz steps by adding the output of a second oscillator in a mixer.

A block diagram of the Collins 51X3 dual-conversion superheterodyne using digital tuning is shown in Fig. 3-12. Twenty-nine crystals cover the band from 108 to 126.9 MHz in 0.1-MHz steps. A local frequency is obtained by doubling the frequency of the third-harmonic crystals and then mixing this signal in the first mixer with the incoming signal. The frequency from the high-frequency oscillator doubler is chosen to be between 17.5 and 18.4 MHz (0.9-MHz spread) below the incoming signal, and this difference frequency is coupled to the second mixer. A digital counter dial with a megahertz knob controls the crystal selected for the input to the first mixer. A 100-kHz (0.1-MHz) knob permits selection of the crystal for the lower-frequency input to the second mixer.

The Collins 618S-4 airborne hf transceiver, shown in the block diagram of Fig. 3-13, is another example of a unit providing frequency synthesis for if stability. In the long-distance hf communication for which this transceiver is intended, single-channel simplex operation is used; that is, both parties receive and transmit on the same frequency. Referring to the block diagram, consider first the transmitting mode. A signal from the 250-kHz crystal oscillator passes to the transmitter mixer, together with a signal from the 144-position crystal oscillator, which has a range of 1.75 to 3.50 MHz. The two signals mix to form a signal of 2.0 to 3.75 MHz; this signal then passes through a variable if amplifier to another transmitter mixer. Its output frequency will be equal to $(F + 250$ kHz$) + NF$, where F is the basic oscillator frequency and N the multiplier.

The signal thus synthesized passes to the final transmitting amplifiers, which are servo-tuned to resonance. Tuning the latter ampli-

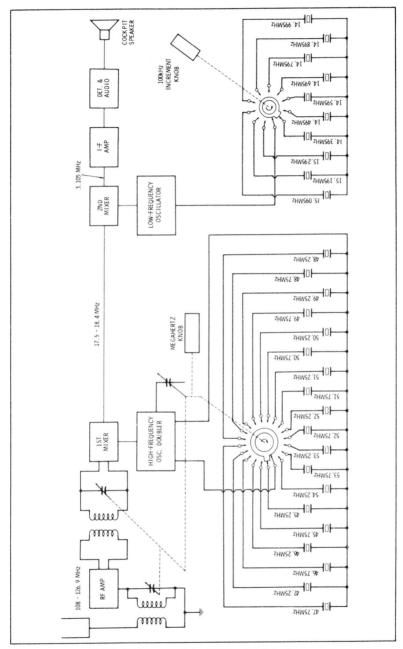

Fig. 3-12. Block diagram of the Collins 51X3 dual-conversion
superheterodyne vhf receiver.

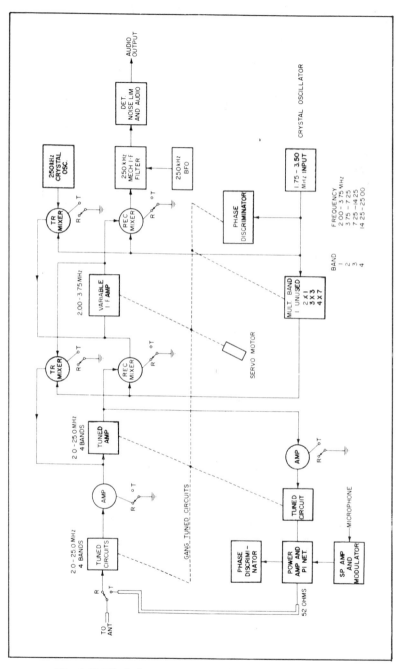

Fig. 3-13. Block diagram of the Collins 618S-4 hf transceiver.

fiers also automatically positions the tuning shafts of the receiving rf amplifier. The receiving system utilizes a 250-kHz if. The signal from the distant station also equals (F + 250 kHz) + NF. Hence, in the receiving mode, a signal comparable to the one from a local oscillator is synthesized by eliminating the 250-kHz oscillator frequency previously added to obtain the transmitter signal. The incoming signal is first directed to a receiver mixer which is also fed the frequency NF, resulting in a mixer output of F + 250 kHz. The signal next passes to another mixer being fed a signal equal to F; thus, the output will be the desired if frequency of 250 kHz.

Phase-Locked Loop Circuitry

Aviation frequencies are assigned in regular 25- or 50-kHz increments and cockpit selection is now made on a digital readout in tens, hundreds, and decimals. The phase-locked loop employing digital circuitry is ideal for this, in addition to the advantages previously mentioned. The system has been highly refined through the advantage of mass-produced amateur- and Citizens-band equipment as well as the aviation market equipment. Both TTL and CMOS specialized IC chips are available. Programmable frequency-counter chips used in this system are an outgrowth of computer development.

Only one crystal is used, which acts as a clock signal that is compared to the phase of a voltage-controlled oscillator (vco). An error voltage is developed that is used to lock the phase of the vco to the clock signal. The versatility of the system comes from the fact that the clock signal can be arranged so that the vco operates over a range of frequencies, each of which is accurately held within the tolerance of the one basic crystal.

A block diagram of a basic PLL system is shown in Fig. 3-14. The phase detector is a relatively simple circuit, contained in an IC package, that compares the phase of two inputs that are of the same frequency and produces a voltage when the two inputs differ in phase. The magnitude of the voltage is a function of the phase difference. The polarity depends upon whether the reference signal leads or lags the other. It will be remembered that:

$$\theta = 2\pi ft$$

where,

θ is phase,
π is 3.14,
f is frequency in hertz,
t is any instant in time.

Thus, when $\theta = \theta'$ there will be no voltage from the phase detector. With a PLL system, however, the vco settles down at a point at which the phase difference produces a voltage that holds the phase, hence, frequency, constant. The frequency at which the vco settles down depends upon the setting of the programmable divider. Its output frequency will be counted down to reach that equal to the stable frequency source. The block diagram is simplified and does not show additional circuitry necessary in the vco. A practical vco can

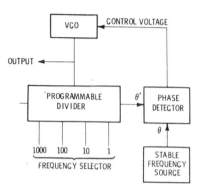

Fig. 3-14. Block diagram of basic phase-locked loop frequency synthesizer.

only operate well over a range of frequencies that may be less than that desired. It is therefore necessary to introduce an adjustment into the vco, such as an offset voltage, that will be made as the frequency selector changes range.

The PLL system just described is suitable at lower frequencies, such as in ADF receivers operating in the 1.0-MHz range. The vhf range, however, is beyond that of IC counters employing TTL or CMOS circuitry. For these higher-frequency ranges, an additional crystal-controlled *offset* oscillator may be employed as shown in Fig. 3-15. This is mixed (or *heterodyned*) with the vco output to form a lower frequency for feeding the programmable divide-

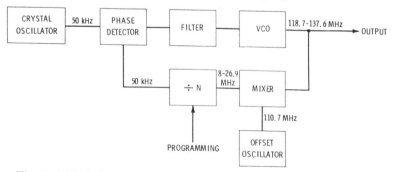

Fig. 3-15. Block diagram of a PLL with output in the vhf range. The offset oscillator brings the vco output down to the operating range of the divide-by-N counter.

by-N counter. The mixer output frequency is the *difference* between that of the vco and the mixer.

As an example of digital frequency synthesis, consider the case where it is desired to synthesize 378 50-kHz channels for an aircraft receiver in the vhf band. Output of the vco will be used as the local oscillator, and the if will be tuned to 10.7 MHz. If the receiver is to be tuned to 108 to 126.9 MHz, the vco output must therefore be greater than the received signal by 10.7 MHz, and the vco range must be 118.7 to 137.6 MHz in 50-kHz increments. We choose an offset frequency of 110.7 MHz (Fig. 3-15). Output of the mixer will then be from 8 to 26.9 MHz (the difference between the offset frequency and the vco frequency). This output is fed to a programmable counter (frequency divider) which counts down the mixer output by N. The value of N will be from 160 to 538 in 538 − 160 = 378 steps, each step being a channel. The reference oscillator is a constant 50 kHz, and the mixer output will always be counted down to this value when the vco stabilizes at the value which is determined by the selection of N.

The programmable counter is a key feature of digital frequency synthesis. This is an integrated circuit that produces a single output pulse for every N input pulses when N is an integer. A typical programmable counter is shown in Fig. 3-16. Note that pins 9, 10, 1, and 15 are the programming inputs. These are each placed at logic

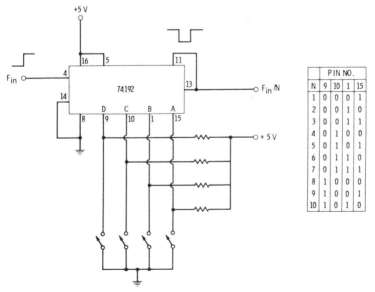

Fig. 3-16. Programmable counter and method of programming. A closed switch = logic 0.

1 or logic 0 to program the counter. For logic 0, the pin is grounded. The table in Fig. 3-16 shows the binary input for the decimal numbers 1 to 10, which sets the divisor, N, into the counter. To set the counter to divide by 6, for example, the switches would be set to logic 0110 (switches for pins numbered 9 and 15 would be closed).

In the synthesizer taken for our example, however, the value of N must be from 160 to 538. This requires using three counters in cascade, as shown in Fig. 3-17. There will be a counter for units, tens, and hundreds. The illustration shows the counter set to divide by 538. Thumbwheel switches are used to set the logic to program the counter from N = 160 to N = 538. Each counter is fed by a switch with 10 positions. The switch automatically removes pins from ground according to the selected value of N. This type of switch is known as a *BCD complement*. It is commonly used for digital inputs.

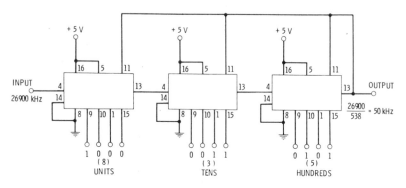

Fig. 3-17. Cascaded programmable counters with logic set for N = 538.

TRANSCEIVERS

Properly speaking, a transceiver is a device which contains circuitry for both receiving and transmitting, but which contains some circuitry common to both functions. This results in weight and cost reduction. To a growing extent, most communication equipment produced for single-engine and light twin-engine airplanes are built as transceivers. Transceivers usually are self-contained with easily accessible controls. In their lightest form, transceivers are available which are completely portable for use in gliders, crop-dusters, or for specialized air-ground communication.

An example of a small portable transceiver is the Radio Systems Technology Model RST-541 two-channel transceiver. This unit is unique in that it is purchased as a kit and may be assembled by the owner. It may not be used, however, until adjusted after assembly

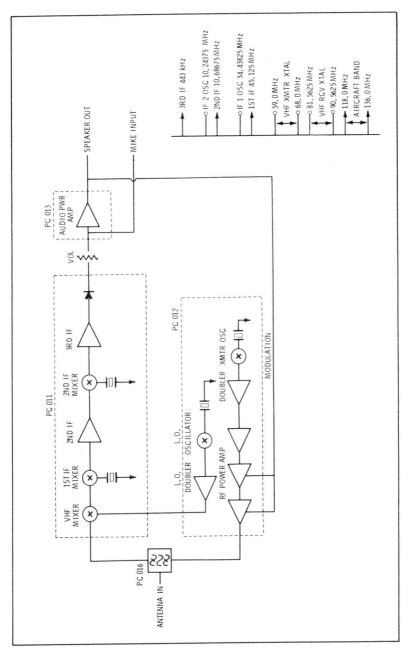

Fig. 3-18. Block diagram of the Radio Systems Technology Model RST-541 two-channel transceiver.

77

by the holder of a First- or Second-Class Radiotelephone Operator's license, who must install the crystals. The assembled unit may be returned to the factory for this purpose. Fig. 3-18 shows a block diagram of the unit. The receiver section uses untuned bandpass-type input with intermediate frequencies of 45.125 MHz, 10.68675 MHz, and 443 kHz. The frequencies have been chosen to prevent spurious local oscillator frequencies from mixing with input signals in the 118- to 136-MHz band, and to provide good image rejection.

Fig. 3-19. The Genave ALPHA/ 100-360 transceiver. (*Courtesy General Aviation Electronics, Inc.*)

The unit is capable of receiving and transmitting only two channels; however, it is a simple example for examining the principle of a transceiver. Notice that the power amplifier used for the receiver audio also acts as the modulator. This is the usual method with transceivers. A relay switches the antenna from receiver to transmitter and enables the transmitter when the microphone button is actuated. When it is desired to switch to the other channel, a simple switching arrangement places the other crystals in the circuit (frequency synthesis is not used since the unit operates on only two channels).

To satisfy requirements for IFR flight, it is necessary to have the capability to transmit and receive at 50-MHz increments throughout the band from 118.0 MHz to 127.9 MHz. The Genave Alpha/100-360 shown in Fig. 3-19 is an example of a transceiver with this capability. The receiver uses double conversion. The unit is shown in block-diagram form in Fig. 3-21 and consists of the following sections:

Receiver.
High-Frequency Oscillator.
Exciter.
Transmitter.
Audio Amplifier/Modulator.
Regulated Power Supply.

The local oscillators, exciter, and transmitter are contained in replaceable shielded modules. A main printed-circuit board contains power supply and audio amplifier/modulator. This transceiver is de-

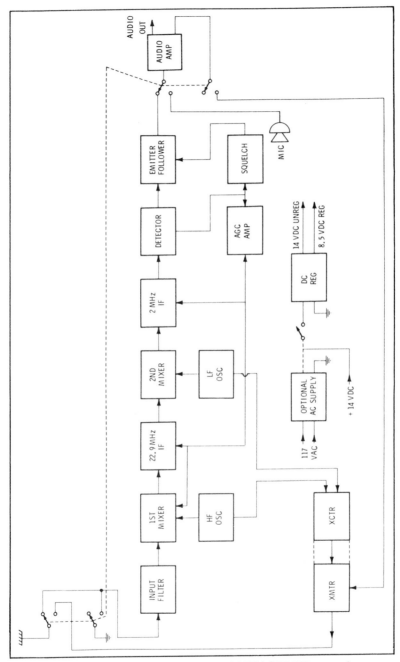

Fig. 3-21. Block diagram of the Genave ALPHA/100-360 transceiver.

signed to be used also as a ground station, and an optional module may be added to the power supply for 117-volts ac operation.

The receiver is of the double conversion type. The first if has a center frequency of 22.9 MHz and the second if operates at 2.0 MHz. The "front end" of the receiver is the first converter. An untuned bandpass filter (Chebyshev type) forms the rf input.

Signals from the antenna are applied to the transmit/receive relay K101 via J102, the antenna jack. When K101 is in the receive position, the signal from the antenna jack will be routed to the input filter. The input filter is a 5-pole Chebyshev filter consisting of L101 through L105 and associated tuning and coupling capacitors. This bandpass filter allows signals between 118.00 and 135.95 MHz to pass to the first mixer.

The high-frequency oscillator output is applied to the base of Q101, the first mixer. This signal is 22.9 MHz above the frequency of the desired incoming signal. The gain of the first mixer is controlled by the agc voltage applied to the base through R101. The 22.9-MHz difference frequency from the first mixer is fed to the first if consisting of Q102 and associated circuitry. The first if has a center frequency of 22.9 MHz and a bandwidth of 1 MHz. This stage is also controlled by the agc voltage that is applied to the base of Q102 through R106.

Output of the first if is applied to the second mixer consisting of Q103 and associated circuitry. Output of the low-frequency oscillator is applied to the base of Q103 through C122. The signal from the low-frequency oscillator is 2 MHz above the desired first if frequency. The resulting 2-MHz difference signal is then fed to the second if. The 3-stage second if consists of Q104, Q105, Q106 and associated circuitry. The first two stages of the second if are agc controlled. This agc voltage is applied to the bases of Q104 and Q105 through R119 and R126, respectively.

Output of the second if is applied to the detector CR102. Audio from the detector is fed to the agc amplifier, audio emitter-follower, and squelch amplifier. The agc amplifier is comprised of Q108 and associated circuitry. The agc voltage, which is fed to the agc amplifier, is determined by back-biasing the detector diode through R137. The varying dc voltage produced is amplified by Q108 and used to control the first mixer, first if amplifier, and second mixer. Components R139, R141, CR103, and C149 form a noise limiter that removes impulse noise from the voice audio. Transistor Q109 and associated circuitry form the squelch amplifier. The squelch amplifier controls the bias level on the emitter of Q107, the voice emitter-follower. Output of the voice emitter-follower is fed to the audio amplifier.

The high-frequency oscillator circuitry consists of Q301, Q302,

Q303, and associated circuitry in a modified Colpitts, crystal-controlled transistor oscillator. Crystals from 69.7185 MHz to 79.2185 MHz in 0.5-MHz steps are used. They are selected by grounding the appropriate crystal lead as selected by the front panel switch, SW101, and the transmit/receive relay.

Output of the oscillator is fed into the base of a Class-AB frequency doubler, Q302. Doubler output is filtered by a 3-pole Chebyshev bandpass filter which reduces spurious levels to 70 dB or more below the reference output frequency. The filter has a passband of 139 MHz to 159 MHz. Filter output is matched to a 50-ohm coaxial cable which is routed to the main circuit board and also to the exciter assembly. The high-frequency oscillator is contained within a separate shielded module.

The low-frequency oscillator uses crystals in the range from 20.487 MHz to 21.437 MHz in 50-kHz steps. The crystals are selected by grounding the appropriate crystal lead as selected by front panel switch SW101.

Transformer T401, used to adjust the low-frequency oscillator, utilizes a pickup link. This link is the first element of a 5-pole Chebyshev low-pass filter consisting of the link on T401, C411, L403, C413, L404, C415, L405, C417, and L406. This filter suppresses unwanted outputs to 70 dB or more below the desired output frequency. The nominal cutoff frequency is 26 MHz. Output of the filter is matched to a 50-ohm coaxial cable that is routed to the main circuit board and then to the exciter assembly. The low-frequency oscillator is contained within a separate shielded compartment of the transceiver.

Inputs from the high- and low-frequency oscillators are fed through resistive attenuators to the balanced mixer consisting of Q501 and Q502. The low-frequency input is applied through a tuned transformer T501, and fed differentially to the transistor bases. The high-frequency input is applied in-phase to both bases. Using this method of feeding the mixer, the high-frequency input, its harmonics, and all even-order harmonics of the low-frequency input are suppressed in the collector circuit. Mixing action occurs in the base-emitter junctions and produces primarily the high-frequency input plus and minus the low-frequency input. Harmonically related spurious outputs also occur, but at lower levels. The desired output frequency is the high-frequency input minus the low-frequency input. The sum and difference frequencies appear in the collector circuit across the primary of T502. A pickup link on T502 forms the first element in a 3-pole Chebyshev bandpass filter consisting of the link on T502, C509, C510, C513, L501, C514, C515, C516, C511, C512, and L502. The nominal bandwidth of this filter is 19 MHz centered around a frequency of 127.95 MHz. This filter suppresses all undesired higher-order outputs of the mixer to 60 dB below the desired output.

The output filter drives an emitter-follower consisting of Q503 and associated circuitry. The emitter-follower drives two common-emitter amplifiers consisting of Q504, Q505, and their associated circuitry. Transistor Q505 feeds the high-frequency filter. The undesired filter is removed from the signal path when +8.5 volts dc is applied to the emitter of the preceding transistor. The frequency selector switch determines which filter will be utilized, applies +8.5 volts dc to the switched A+ line corresponding to this amplifier, and shuts off the transistor.

The low-frequency filter is comprised of L503, L505, L506, and corresponding tuning and coupling capacitors. The low filter has a passband from 118.00 to 127.95 MHz and suppresses all lower-order responses 70 dB or more below the desired output. When output frequencies from 118.00 MHz to 127.95 MHz are desired, this filter is utilized. The high filter has a passband from 128.00 to 135.95 MHz and is comprised of L504, L507, L508, and associated tuning and coupling capacitors. When output frequencies from 128.00 to 135.95 MHz are desired this filter is used.

Outputs of these two filters are combined by a diplexer coil, L509. The summed output is taken from the center-tap of L509 and fed to J501 via C550. The exciter assembly is contained within a separate shielded module that prevents radiation. Drive from the exciter is applied to Q506, a single-tuned bandpass amplifier. At the output of this amplifier, all undesired outputs are more than 70 dB below the desired output.

Output from Q506 is fed to a single-tuned Class-C driver, Q507. The signal from Q508 is matched into the input of Q509 with a split inductor "pi" matching section, consisting of Z503, C566, C567, and Z504. Transistor Q509 is the final power amplifier stage. It is single-tuned into the 6-pole Chebyshev low-pass filter. The function of this filter is to remove harmonics of the output frequency which are generated in the Class-C amplifier stages. The filter reduces harmonics and spurious outputs to 60 dB or more below the desired output. Output of the transmitter connects to the antenna switching section of the transmit/receive relay, K101.

The audio amplifier/modulator is a five-stage direct-coupled Class-B complementary symmetry amplifier consisting of Q110, Q111, Q112, Q113, Q114, Q115, and Q116. The amplifier is provided with ac decoupling and bias stabilization by means of R151, R154, and C157. Capacitors C158 and C159 are used to provide closed-loop stability. High-frequency band shape and rolloff are controlled by the ac feedback network consisting of R162, R163, and C162.

During transmit, low-frequency rolloff is controlled by R149 and

C154. Transformer T113 is used to match impedance and voltage levels to properly modulate the transmitter and speaker. The output voltage is slightly less than 12 volts, thus assuring that the transmitter cannot be overmodulated. The Class-B modulation technique assures that the modulation voltages cannot exceed the voltage applied to the transmitter. Resistor R150 provides a noise-free regulated current to the microphone element. It may be changed in the field to provide proper modulation percentage with nonstandard, low, or high output microphones, such as transistorized types designed for direct replacement of the carbon types. The up modulation, as previously mentioned, is controlled by the Class-B modulation technique. Modulation voltage levels are selected so as to prevent carrier cutoff. This combination limits the modulation to 95% of maximum.

The audio amplifier/modulator circuitry is located on the main circuit board.

Circuits are operated from a regulated power supply consisting of Q117, Q118, and Q119. Zener CR107 determines the necessary reference voltage on the base of Q118. Output of the regulator, nominally 8.5 volts, is set by R169 which determines the bias level on the base of Q119. The differential amplifier formed by Q118 and Q119 is used to control the series regulator transistor Q117 via the collector-to-base connection of Q118 and Q117. Resistor R165 supplies a portion of the load current that allows Q117 to operate within dissipation capabilities.

MAINTAINING AND TROUBLESHOOTING COMMUNICATIONS SYSTEMS

The usual pilots' complaints of trouble in communications equipment are as follows:

Total failure of transmitter and/or receiver.
Failure of receiver or transmitter on one channel only.
Noisy reception.
Weak reception.
Garbled or distorted reception.
Garbled or distorted transmissions as reported by other stations.
Any of the foregoing on an intermittent basis.

In troubleshooting a radio system, a good rule is to perform the easiest check first. Thus, unless the complaint involves an obvious trouble within the unit itself—for example, failure on only one channel—the first checks should be external. The antenna (including the lead and connections), fuses, power cables and their connections are some of the things that should be suspected. Much depends, of course, on the individual system.

Noisy reception, if interpreted to mean signals obscured by background noise, is usually traceable to the antenna and associated input circuitry. Weak reception, in the absence of noise, usually means poor audio- or squelch-circuit performance.

In reports of complete receiver failure, the squelch-control setting should be checked as a matter of course. Garbled or distorted reception can generally be traced to a defect in the audio circuitry, with the speaker also as a possible cause. If a speaker is suspected, a quick switch to the earphones will serve as a check. Once it is determined that the radio equipment rather than its interconnections or accessories is at fault, troubleshooting on the bench is the next step.

Some of the more common causes of trouble in radio equipment— defective transistors, connectors, etc.—are not too difficult to isolate. Capacitors, resistors, coils, and other components also become defective but are usually more difficult to track down. Most troubleshooting in receivers can be performed with the aid of a volt-ohm-milliammeter and the proper signal generator.

If the gain of a receiver seems low, it can be checked with a suitable signal generator and power-output meter. The gain is determined by the amount of rf input signal required to produce the maximum rated output power (indicated by the meter) of the receiver. Gain then is the ratio of output to input and is usually expressed in decibels (dB).

If voltage measurements appear correct, and no other trouble is obvious, low gain may mean that realignment is required. Usually, rf alignment is not necessary unless tuning capacitors, rf coils, and similar components have been replaced. In high-frequency if circuits, even the replacement of a transistor may detune the if transformer, because of a difference in the input capacitance of the replacement transistor. The rf stages are "peaked" by the methods given by the manufacturer, who usually specifies a sweep generator.

Many tests can be performed on the transmitter without removing it from the airplane. The signal strength, for example, can be checked with a field-strength meter. A small antenna on the instrument picks up transmitter energy near the airplane. This will immediately indicate whether or not a transmitter signal is being radiated. The meter deflects in proportion to the actual field strength at a given point. The actual rf output power (wattage) of the transmitter can be measured by feeding the rf energy into a dummy load. To meet RTCA requirements, the dummy load used for transmitter testing should have a resistance within 10%, and a reactance of not more than 10%, of the characteristic impedance of the transmission line.

Based upon these conditions, the power output of a transmitter will be:

$$P = E_o^2/R_o \qquad \text{(Eq. 3-7)}$$

where,
E_o is the rf voltage across the dummy load,
R_o is the characteristic impedance of the antenna transmission line.

A simpler way to determine transmitter output power is by using either an in-line or a terminating-type rf wattmeter. Power can then be read directly from the meter without computation. The in-line, or series-type, meter requires some form of termination, which can be the antenna itself or a separate dummy load. The terminating-type rf wattmeter, on the other hand, requires no termination since it is designed to serve as a dummy load as well.

Fig. 3-22. The IFR, Inc., Model FM/AM–1000 S allows thorough check of a communication system from the ramp, or it may be used on the bench. (*Courtesy IFR, Inc.*)

IFR, Inc., of Wichita, Kansas, produces a completely portable battery- or ac-powered test set for testing communication equipment. The Model FM/AM–1000 S (Fig. 3-22) consists of a calibrated rf source as well as a built-in oscilloscope which also becomes part of a spectrum analyzer capability. The internal battery allows at least 40 minutes of full operation, and automatically recharges when connected to an ac line.

Transmitter monitoring functions (frequency error, deviation, power, etc.) are displayed on front-panel meters or the oscilloscope, which doubles as a general-purpose service tool. Accurate frequency measurements are made possible by the over-stabilized master oscillator which is referenced to WWV during normal operation of the service monitor.

Internal modulation in 0.1-Hz increments to 10 kHz is provided by a frequency-synthesized tone generator to enable subaudible tone squelch and tone-selective signaling tests. A front-panel connector provides modulation tone output. Low-frequency fm response permits swept receiver if alignments with the oscilloscope. The generator can also be externally modulated down to 2 Hz for testing data transmission systems. A built-in beat-frequency oscillator permits single-sideband and cw monitoring.

The FM/AM–1000 S contains a versatile spectrum analyzer in the oscilloscope module. This permits off-the-air monitoring of transmitter spectrums, as well as providing a general-purpose spectrum analyzer/oscilloscope for general shop use. The FM/AM–1000 S uses a temperature-compensated crystal oscillator (TCXO). The TCXO does not require warm-up time to stabilize, making the unit ready to use as soon as power is applied.

4

VHF Omnirange

The old four-course, low-frequency range stations were inflexible and subject to atmospheric noise under crucial bad weather conditions. Older pilots who have flown an airplane on instruments in bad weather, with static crackling above the constant 1020-Hz tone, agree that this method does not meet the requirements of today's navigation.

Near the end of World War II, it became obvious that a more effective means was needed to guide aircraft in flight from point to point. By 1940, the CAA (now FAA) had begun the development of what is now known as the standard vhf phase-comparison omnirange. The system is sometimes called merely *omni* or *VOR*, which stands for Very-high-frequency *Omni*Range.

In 1945, the CAA system was presented to the Radio Technical Commission for Aeronautics, and technical standards were discussed. Shortly thereafter, the United States government offered the omnirange system to the world, and subsequently it was adopted as part of a standard world airways system.

VOR stations appear on United States Aeronautical Charts as a compass circle centered over the station (see Fig. 4-1). As a convenience to the pilot or navigator, the VOR station is always oriented relative to magnetic north, with local corrections included. To determine the bearing of an aircraft from a station by using a manual omni system, the pilot first tunes to the station and then manually turns the omnibearing selector (OBS) until it nulls. The reading on the OBS will then be the bearing to or from the station. Regardless of the heading of the airplane itself, it will at all times be some-

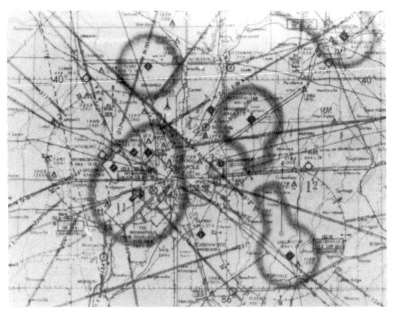

(A) *Segment of a U.S. Sectional Aeronautical Chart (VFR).*

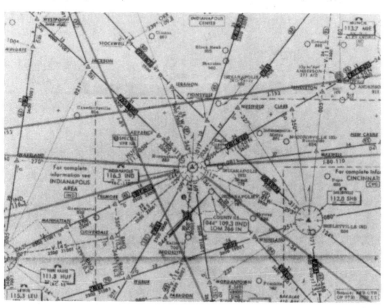

(B) *Segment of a U.S. Low-Altitude En-Route Chart (IFR).*

Fig. 4-1. United States Aeronautical Charts.

where along the radial indicated by the OBS. In other words, at some distance from the station, an airplane could fly tight circles over a point without disturbing the null of the VOR steering needle.

A sense indicator ("To-From" indicator) is provided for determining whether the indicated bearing is "To" or "From" a station. If the bearing is indicated as *to* a station and the pilot desires to "home" on the station, he turns the airplane to the heading indicated. The airplane is then flown to keep the VOR needle centered. If there is a crosswind, it will be necessary to hold a different heading from that indicated on the OBS in order to cause the airplane to progress along a VOR radial. More expensive VOR systems employ a servo loop to keep the VOR indicator always centered on the correct radial. In these systems, no action on the part of the pilot is required.

The location of an airplane can be determined by tuning to two VOR stations in succession, determining each bearing, and then drawing an intersection. Actually, the FAA has set up fixed VOR intersections along the airways. These serve as traffic control points. Aircraft en route on IFR flight plans are required to report, hold, change altitude, or change route over these VOR intersections, which are shown on all radio facility charts.

The VOR may be used for landing approaches to airports that have been authorized by the FAA. Such approaches, however, must be carried out with higher minimum ceilings than ILS precision approaches (see Chapter 6). The approach is made along a radial that is in line with the runway. The station may be a TVOR located on the airfield, or it may be a more distant en-route VOR located in line with the runway. A VOR approach is begun by flying outbound over the VOR station, then making a 180° "procedure turn," then returning over the station, reporting to ATC when over the station, and beginning the letdown. Only the altimeter is available for altitude information. The aircraft must break out under the overcast when still high enough to allow for possible altitude error in the approach.

As the airplane proceeds along a VOR radial, especially when making a VOR approach, crosswind allowance must be determined. When the airplane is flown on the same heading as the VOR radial but the needle refuses to remain centered, this indicates that a crosswind component is causing drift. Crosswind or drift allowance is determined by a process known as *bracketing*. The pilot makes a heading correction in a direction that will center the needle. If the needle then centers, but continues to deviate in the opposite direction, another heading change is made in the opposite direction, but at only half the angle. This process is continued until a heading is found that holds the airplane on the radial.

PRINCIPLE OF OMNIRANGE

To grasp the principle of omnirange, imagine a sharply focused beacon light arranged to rotate at a constant speed. Whenever the beacon points exactly north during its rotation, another light is arranged to instantaneously flash in all directions. By knowing the time required for a full rotation of the beacon and comparing it with the flashes observed from the fixed light when the beacon has reached its northerly point, an observer can determine his direction from the beacon.

The same principle can be applied to radio emissions. An omni ground station consists of an antenna arranged to rotate a directional beam 1800 revolutions per minute, or 30 per second. Thus, an observer located at any point will receive a signal 30 times per second, and it will appear as 30-Hz amplitude modulation. Another antenna radiates a signal in all directions to provide a reference signal. Radiation from the latter antenna contains a 30-Hz reference signal so arranged that it will appear in phase with the signal from the rotating beam when the observer is north of the station. Hence, the relationship between the two signals received will be different for each point of the compass.

The 30-Hz reference signal must be radiated in some manner that will not interfere with the rotating pattern signal. This is done by frequency-modulating a 9960-Hz subcarrier at the 30-Hz reference rate, which is carefully synchronized with the rotation of the direction beam. Signals at the ground station are continuously monitored by automatic means, and the station is taken off the air if any condition develops that would mislead a pilot.

A vhf receiver in the airplane receives the sum of all modulation components from the ground station. A filter separates the 30-Hz am component of the rotating beam from other modulation components. Likewise, a filter selects the 9960-Hz fm component. The fm signal is then detected to recover the 30-Hz reference signal, which in turn is compared in phase with the signal from the rotating beam.

The sum of all modulation components from an omnirange station results in a maximum of 90% modulation. The rotating pattern is adjusted to produce the effect of 30% modulation. The reference subcarrier is also adjusted to produce 30% modulation. This leaves an additional 30% of modulation power, which is used for voice modulation. Voice transmissions are used for identifying the particular station, for weather broadcasts, and also for communication with en-route aircraft, which can be carried out without affecting the performance of the VOR as a navigational device.

A typical airplane receiver for omnirange navigation will exhibit a sensitivity such that a 3-microvolt input signal modulated at 30% with a 1000-Hz tone will produce 200 milliwatts into a load with a 6-dB signal-to-noise ratio. The agc is incorporated to maintain not more than a 3-dB variation over an input range of 5 to 50,000 millivolts. Typical audio fidelity is 6 dB between 300 and 3500 Hz. Almost all navigation receivers tune the entire aircraft band and therefore can also be used for communications. A squelch circuit is incorporated, but is automatically disabled in some receivers when they are tuned below 118 MHz. The reason is that pilots may inadvertently set the squelch too high and thus disable the receiver for navigational reception. Less expensive receivers are continuously tuned, but those preferred for regular IFR service are digitally tuned utilizing crystals.

OMNIBEARING INSTRUMENTATION UNITS

Actual VOR signal demodulation and phase comparison are usually carried out in a separate unit from the receiver. Fig. 4-2 shows a block diagram of a simple omnibearing instrument. Audio output from the receiver is fed to two filters in the instrument. One filter has a passband of 9480 to 10,440 Hz and therefore separates the fm reference signal; the other filter passes only the 30-Hz signal which results from the rotating pattern of the distant station. The fm audio-reference signal is first limited and then fed to a discriminator detector, resulting in a 30-Hz signal. The signal from the low-pass filter passes to a phase shifter (usually a resolver), and from there both signals pass to a phase detector.

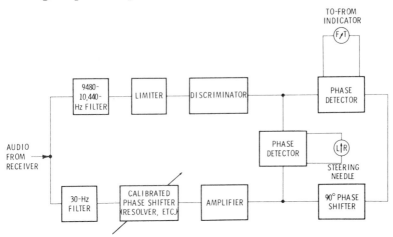

Fig. 4-2. Block diagram of a typical omnibearing instrument—also known as a VOR instrumentation unit, converter indicator, etc.

The phase detector actuates a galvanometer movement visible to the pilot. The polarity of the galvanometer is so chosen that when the needle is centered, the setting of the phase shifter (resolver) will indicate the bearing of the VOR station. To fly on this bearing, the pilot keeps the needle centered by turning *toward* it when flying:

1. On a heading *to* a station and the OBS is set on the heading *to* the station.
2. On a heading *from* a station and the OBS is set on the heading *from* the station.

Conversely, when flying on a heading *to* a station and the OBS is on the heading *from* the station (or *from* a station and the OBS heading is *to* the station), the pilot must turn *away* from the needle to keep it centered and stay on course.

In a phase detector, there are two points 180° apart where a null can be reached. Therefore, an airplane may be either on the bearing or its reciprocal when the steering needle has been nulled. In one case, the reference voltage lags the bearing signal (when flying *from* a station), and in the other it leads (when flying *to* a station).

To determine whether a bearing is *to* or *from* a station, a fixed 90° phase shift is applied to the reference voltage, and the voltage again compared with the signal by using another phase detector. Fig. 4-3 shows the phase relation arising from the "To-From" bearings. Only when an aircraft is on a line through the station perpendicular to the bearing set on the OBS will the "To-From" indicator read zero. The "To-From" indicator also is a galvanometer type of movement. In place of a needle, most movements are in the

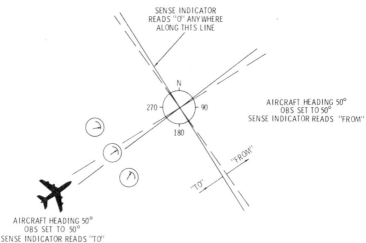

Fig. 4-3. The phase relation arising from the "To-From" bearings.

form of a light metal "flag" moving behind a cutout on the face of the instrument.

The omni station antenna is designed so that there is no radiation directly above the station. Nevertheless, this space will contain a small amount of energy which has been reflected from the station structure, nearby trees, and other objects. The resulting signal will be directionally incoherent and since the field strength is weak, the agc of the receiver will greatly increase the sensitivity, causing erratic motion of the steering needle. The space thus affected above the station is called the *cone of confusion*. Over this cone, the steering needle will suddenly swing in one direction and then will deflect to the opposite while the "To-From" flag moves from one position to the other. The cone of confusion is used as a marker point, and pilots are often requested to report when passing over it.

Fig. 4-4. The Collins 344D-2 is an example of a lightweight VOR unit designed for mounting on the instrument panel of a small airplane. (*Courtesy Collins Div., Rockwell International*)

Fig. 4-4 shows an example of a lightweight, compact VOR unit that will easily fit into the instrument panel of an airplane.

SOURCES OF ERROR IN VOR

The accuracy of VOR varies from ±2° in the better units to over ±4° in the less expensive units. For IFR flight, an accuracy of at least ±4° is mandatory when the accuracy check is made at a certified ground checkpoint. When the accuracy check is made over a certified airborne checkpoint, the accuracy must be within ±6°. VOR units with less accuracy are still satisfactory to the VFR pilot, however. The accuracy, reliability, and cost of each system are closely interrelated.

Errors in the overall VOR system are classified as *instrument errors* and *propagational errors*. Propagational errors are:

Vertical pattern effects.
Site effects.
Terrain effects.

Vertical pattern effects have to do with the nature of waves emitted from above the antenna ground plane. There are two waves emitted by the VOR station. A direct wave travels from the antenna directly to the airplane. A second wave is reflected from the ground plane and reaches the airplane after reflection. The overall strength of the wave reaching the airplane will be the algebraic sum of these two waves. At certain points in space the two waves will be additive, at other points subtractive. A vertical radiation pattern of the wave will appear "scalloped," that is, at certain angular heights, signal strength will be lower.

Site errors are caused by reflection from terrain near the VOR station. Terrain errors are caused by reflections from more distant terrain features. The effects of both site and terrain error are to displace the apparent location of the station, or at least cause the radials to be unevenly spaced. In general, the FAA tries to site VOR stations so that for a radius of 1000 feet around the station the terrain will be both clear and flat.

Instrument errors have to do with the equipment itself. These are:

Ground station octantal error.
Receiver indicational error.
Polarization error.

Octantal error is caused by the failure of the antenna system to emit with constant directional characteristics as the beam is moved through 360°. Receiver indicational error originates at the receiver and OBS. Polarization error takes place when the receiver reacts to the small portion of waves that are vertically polarized. This occurs when the airplane banks and tilts the VOR antenna so that it receives a portion of the vertically polarized wave. Polarization errors are greater at higher angular heights from the station.

The effect of an instrumental error on aircraft course is to displace the aircraft to another location from the selected VOR radial. Fig. 4-5 shows this effect. The true position of the aircraft from the nearest point on the selected radial is measured from a right angle drawn to the selected radial, as shown. The error distance, d, is D sin e, where e is the angular instrumentation error and D is the distance to the station.

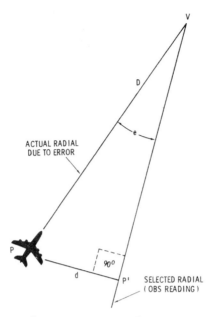

Fig. 4-5. The effect of an instrumental error on aircraft course.

VOR stations are calibrated and checked by specially equipped FAA airplanes and other means. Of the errors mentioned, the aircraft operator can only control receiver error.

MANDATORY VOR TESTS

Federal Air Regulations require a pilot to personally check VOR. Requirements for mandatory VOR checks are given in Part 91.25 as follows:

91.25 VOR EQUIPMENT CHECK FOR IFR OPERATIONS.

(a) No person may operate a civil aircraft under IFR using the VOR system of radio navigation unless the VOR equipment of that aircraft—

(1) Is maintained, checked, and inspected under an appropriate procedure; or

(2) Has been operationally checked within the preceding ten hours of flight time and within ten days before flight, and was found to be within the limits of the permissible indicated bearing error set forth in paragraph (b) or (c) of this section.

(b) Except as provided in paragraph (c) of this section, each person conducting a VOR check under paragraph (a) (2) of this section shall—

(1) Use, at the airport of intended departure, an FAA operated or approved test signal or a test signal radiated by certificated and appropriately rated radio repair station or, outside the United States, a test signal oper-

ated or approved by appropriate authority, to check the VOR equipment (the maximum permissible indicated bearing error is plus or minus 4 degrees);

(2) If a test signal is not available at the airport of intended departure, use a point on an airport surface designated as a VOR system checkpoint by the Administrator or, outside the United States, by appropriate authority (the maximum permissible bearing error is plus or minus 4 degrees);

(3) If neither a test signal nor a designated checkpoint on the surface is available, use an airborne checkpoint designated by the Administrator or, outside the United States, by appropriate authority (the maximum permissible bearing error is plus or minus 6 degrees); or

(c) If dual system VOR (units independent of each other except for the antenna) is installed in the aircraft, the person checking the equipment may check one system against the other in place of the check procedures specified in paragraph (b) of this section. He shall tune both systems to the same VOR ground facility and note the indicated bearings to that station. The maximum permissible variation between the two indicated bearings is 4 degrees.

(d) Each person making the VOR operation check as specified in paragraph (b) or (c) of this section shall enter the date, place, bearing error, and his signature in the aircraft log or other permanent record. In addition, if a test signal radiated by a repair station, as specified in paragraph (b)(1) of this section, is used, an entry must be made in the aircraft log or other permanent record by the repair station certificate holder or his representative certifying to the bearing transmitted by the repair station for the check and the date of transmission.

TESTING AND REPAIRING VOR EQUIPMENT

Approved testing of VOR equipment requires that some accurate means of simulating a VOR signal be at hand; and for this purpose a special signal generator is required. Some manufacturers have developed such signal generators as well as other special VOR test equipment. Fig. 4-6 shows an example of such a generator, the Type H-14A produced by Aircraft Radio Corporation. This instrument, when connected to an antenna, is capable of radiating a low-powered signal usable for checking VOR systems in airplanes on the ramp, or in flight within a radius of about five miles. Means are also provided for transmitting instructions to pilots, using the test frequency. For bench testing, the H-14A provides signals of accurately known frequency and amplitude, with VOR modulation that can be used for quantitative as well as functional tests. A block diagram of this signal generator is shown in Fig. 4-7.

Two optional crystal-controlled channels are provided by the H-14A. These can be anywhere from 108 to 132 MHz, but 110.9 and 114.9 are preferred. A 9960-Hz subcarrier with a 30-Hz frequency-modulation reference, and a 30-Hz am signal that can be varied in steps of 15° with respect to the reference phase, are provided. The percentage of modulation may be adjusted separately for each component. A calibrated attenuator regulates the carrier output, and up to 200 feet of RG-58/U coaxial cable can be used to feed a suitably located antenna. During checks in flight, the signal can be identified by voice modulation, or by turning the IDENTIFIER switch on (which causes continuous 1000-Hz dashes to be transmitted). In addition to testing VOR, the H-14A can simulate the ILS localizer signal.

The VOR test signals are generated by tone wheels driven by a synchronous motor that is energized from a 115-volt, 60-Hz source. The instrument therefore must be operated from a stabilized 60-Hz supply. Four coils are located around the tone wheel at precise angular spacing of 0°, 30°, 120°, and 180°. These are connected to the omni track switch that selects pairs of coils in various combinations.

Fig. 4-6. The ARC Type H-14A signal generator for checking VOR equipment. (*Courtesy Aircraft Radio Corp.*)

The Tel-Instrument Electronics Corporation Model T-12A Nav-com Generator shown in Fig. 4-8 is a modern universal NAV/COM test set that supplies VOR, LOC, glide slope, marker beacon, and communication test impulses. All signals are generated with solid-state circuitry. An advantage of this instrument is that all transistors operate between saturation and cutoff, and no amplification is used. A tightly regulated power supply thus ensures constant levels within the circuitry.

The instrument supplies a standard variable VOR signal. For testing glide slope and LOC channels, 150- and 90-Hz signals are supplied that are adjustable from −4 to +4 dB. For testing the marker-beacon receiver, the generator supplies 400 Hz, 1300 Hz, and 3000

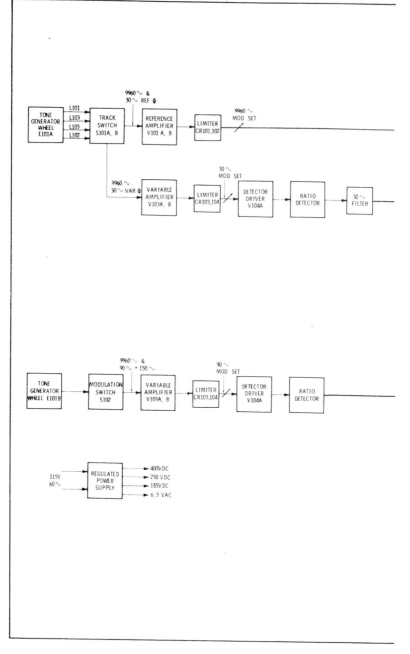

Fig. 4-7. Block diagram of the

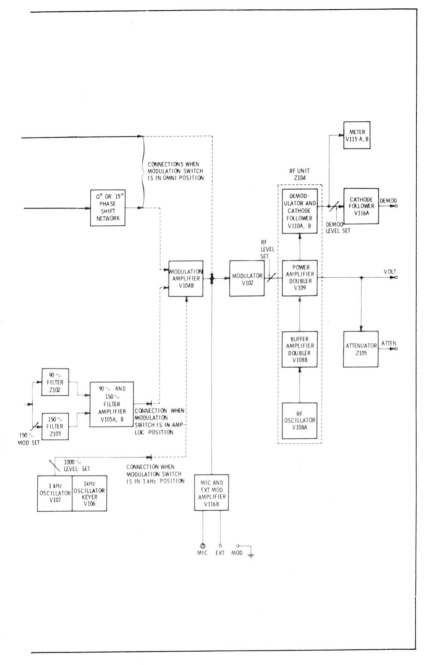

ARC Type H-14A signal generator.

(A) The T-12A used in combination with the T-12-1A LOC/MB/VOR
R.F. Head.

(B) The T-12-2A Glide Slope Head that can also be used with the T-12A.

Fig. 4-8. The TIC Type T-12A Navcom Generator.
(*Courtesy Tel-Instrument Electronics Corp.*)

Hz, which may be keyed at a 25-Hz rate. A 1020-Hz identifier tone is available for mixing with both VOR and LOC signals, and a microphone input is available for mixing voice with the test signals.

When used in combination with the T-12-1A LOC/MB/VOR R. F. Head (as shown in Fig. 4-8A), the combination supplies signals for localizer, marker beacon, and VOR receivers. When used with the Type T-12-2A Glide Slope Head (Fig. 4-8B), the T-12-2A pro-

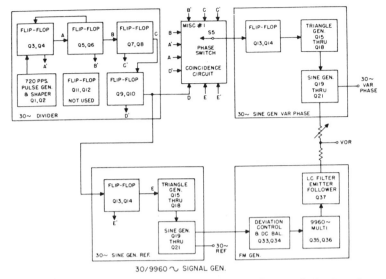

Fig. 4-9. Block diagram of the TIC Type T-12A showing derivation of the VOR signal. (*Courtesy Tel-Instrument Electronics Corp.*)

vides up to 20 crystal-controlled frequencies from 329 MHz to 335 MHz in 300-kHz increments, modulated to appear as a glide-slope signal.

The Tel-Instrument T-12A generates the various tones using solid-state flip-flops in combination with unijunction relaxation oscillators and other pulse circuitry. The VOR signal is derived as follows (see Fig. 4-9):

A basic 720 Hz is generated by a stable relaxation oscillator, then divided down to 30 Hz by a string of five bistable multivibrators. The reference 30-Hz sine-wave generator consists of a square-wave generator, a shaper that changes the square wave to a triangular wave, and a diode function generator that converts the triangular wave into a square wave. The square-wave generator is a bistable multivibrator. Trigger for the multivibrator is the 30-Hz signal derived from the frequency divider.

The square wave from the 30-Hz reference phase generator is fed through the phase switch, thus gating the square-wave generator of the variable-phase, sine-wave generator. When the phase switch goes from 165° to 180°, this gate is inverted so that the switch is capable of varying the phase from 180° to 345° in 15° steps. By an arrangement of charging and discharging capacitors, the triangular wave is converted to a sine wave of negligible distortion.

The fixed-phase, 30-Hz sine wave is applied to the bases of the free-running multivibrators, Q35 and Q36. The 5000-ohm fre-

101

quency-centering control is adjusted so that the multivibrator varies ±480 Hz on either side of 9960 Hz. The frequency-modulated square wave is then filtered through a low-pass LC filter.

As an example, the directions for use of the T-12A in performing localizer and omni alignment on the Genave ALPHA/200 NAV/ COM transceiver are as follows:[*]

Localizer Alignment

1. Connect the receiver to the alignment and test setup shown in Fig. 4-10.
2. Set the Omni/Localizer Simulator to a convenient localizer frequency. Do not use an omni frequency.

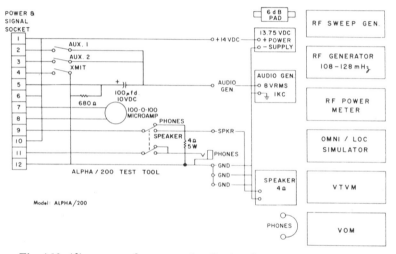

Fig. 4-10. Alignment and test setup for aligning the Genave ALPHA/200.
(*Courtesy General Aviation Electronics, Inc.*)

3. Adjust the rf output of the simulator to 500 microvolts. Set the modulation control for a centering signal. Adjust the modulation of the carrier to the proper level.
4. Turn the receiver on. Set the NAV/COM switch to NAV. Tune the radio to the frequency selected in Step 2. The "TO" lamp should light.
5. Adjust R237, LOCALIZER BALANCE, for a centered meter.
6. Set the modulation control on the generator to +4 dB. The meter should deflect approximately to the edge of the blue color band. There is no adjustment for deflection sensitivity. A deflec-

---

[*]See *ALPHA/200 NAV/COM Maintenance Manual*, General Aviation Electronics, Inc., Indianapolis, Ind., 1968.

tion within $-\frac{1}{8}$ inch to $+\frac{1}{4}$ inch of the color-band edge is normal.

7. Set the modulation control on the generator to −4 dB. The meter should deflect approximately to the edge of the yellow color band. The same comments as in Step 6 apply.

Omni Alignment

1. Connect the receiver to the alignment and test setup shown in Fig. 4-10.

2. Set the Omni/Localizer Simulator to a convenient omni frequency. Do not use a localizer frequency. Adjust the rf output to 500 microvolts. Set the course selector for omni modulation at 90°. Adjust the modulation of the carrier to the proper level.

3. Connect an ohmmeter between the yellow wire terminal and the gray wire terminal on R213, the OBS potentiometer. Set the ohmmeter to R × 1. Adjust the OBS control for a minimum resistance reading. The resistance will be less than 10 ohms. The minimum should occur within $\frac{1}{2}$° of 90° indicated on the OBS dial. If it is further off than this, loosen the set screw in the collar of the OBS drum and set the dial to 90° with a minimum resistance reading on the ohmmeter. Tighten the set screw. Disconnect the ohmmeter.

4. Turn on the receiver. Set the NAV/COM switch to NAV. Tune the radio to the frequency selected in Step 2. Set the OBS dial on the receiver to 90°.

5. Delete the 9960-Hz modulation of the carrier. Adjust R259, AM BALANCE, for a centered meter.

6. Delete the 30-Hz modulation from the carrier and apply the 9960 Hz. Adjust R252, FM BALANCE, for a centered meter.

7. Rotate the OBS dial on the receiver to 270°. Readjust the FM BALANCE control for $\frac{1}{2}$ of the indicated error, if any.

8. Apply both 30-Hz, and 9960-Hz modulation to the carrier. Set the receiver OBS dial to 90°. Adjust R230, PHASE CORRECT, for a centered meter. The "TO" light should be on.

9. Set the simulator course selector to 270°. Readjust the PHASE CORRECT control for $\frac{1}{2}$ of the indicated error, if any. The "FROM" light should be on.

10. Set the simulator course selector and the receiver OBS dial to 0°. Adjust R209, PHASE SHIFT, for a centered meter. The "TO" light should be on.

11. Set the simulator course selector to 180°. Readjust the PHASE SHIFT control for $\frac{1}{2}$ of the indicated error, if any. The "FROM" light should be on.

12. Check the accuracy of the omni at the cardinal points. The accuracy should be within ±2°. An error greater than this at

any one point can be reduced by "trimming" the PHASE COR-RECT or PHASE SHIFT controls.

Troubleshooting any VOR system is comparatively simple, provided the technician understands the principles of omnirange. The first step is to ascertain whether the receiver is at fault, which, generally speaking, is confirmed by the lack of omni indication. Usually, loss of either 30-Hz signal deactivates the needle. If the instrument shows large variable errors throughout 360°, the trouble is usually in the 30-Hz channel—including the variable phase shifter, be it a resolver or whatever.

(Important: Whenever *any* component of the VOR circuit is replaced, a complete accuracy test must be run.)

5

Navigation Systems

Navigation is the process of analyzing and predicting relative motion in order to enable movement from one point to another. Motion is comprised of speed and direction. These two parameters must be known with as much accuracy as is economically possible.

Motion can be measured by various means. The most exotic navigational systems make use of inertial forces. Inertial forces relate to coordinates in free space itself. Thus, the precession of a gyroscope (Chapter 11) can be used as a means of navigation. This is the principle of *inertial navigation.* Inertial navigation systems are expensive and mainly used in military or intercontinental commercial aircraft.

Radio waves offer the most practical means of navigation now in use. ADF, VOR, DME, and ILS are established examples of radio navigation. The most recent developments in radio navigation use miniature on-board computers to analyze aircraft motion relative to points fixed by radio. From this, the present position of the aircraft may be computed on a continuous basis.

Dead reckoning is the process of pointing a vehicle in a known direction and proceeding at a known speed to arrive at a point. All navigation begins with this. Once leaving the ground, however, the airplane moves relative to air motion, or wind. Obviously, this introduces variation in the flight path relative to ground. It is not practical to accurately measure the relative velocity of the wind at all points of the flight. This causes the dead-reckoning position to vary from the true position relative to the earth, even though an attempt is made to compensate for wind direction and velocity. It is also not possible to know the aircraft velocity through the air (airspeed) with absolute accuracy.

Navigation can be considered as a dead-reckoning system with constant "updates" from some means. This is a general concept that even applies when the aircraft is flown along a reference beam such as a localizer. Chapter 12 analyzes how a pilot constantly makes corrections (updates) in the path of flight using a flight director, or how an autopilot becomes part of a flight servo loop. The accuracy of the flight path depends upon the frequency and accuracy of the updating process.

The central element of the navigational process is a computer. This may be a slide rule, such as the traditional "E6-B," or a highly accurate electronic digital computer. The computer solves the dead-reckoning problem. Using the present position, airspeed, and wind component, the computer determines the distance, time, and heading to reach a given point. Updates may be determined in several ways, such as from a VOR "fix" (the intersection of two VOR radials), or simply by visual means when passing over a known location. The most sophisticated navigation systems combine inertial input, star fixes, and radio fixes such as LORAN, as well as VOR, ADF, etc., which supply continuous updating. When an update is made, it is possible to work back to determine the true wind and airspeed by computing the amount of correction from the last update. A correction is then made in computing a new heading and time either to destination or "waypoint."

A flight is planned by a series of waypoints. The waypoint is a point between two headings along the path to ultimate destination. Computations are made to arrive at the heading and time to reach the next waypoint. Waypoints are chosen based upon routing and conditions of the flight. On a long direct flight between two points where wind and airspeed are fairly constant, waypoints may be fewer and farther apart.

A navigation system may be programmed with waypoints, VOR frequencies, and other information prior to flight. The flight may then progress according to the flight plan with the system providing the course automatically.

PROGRAMMABLE CONTROL

Advances in solid state technology have made it possible to incorporate thousands of transistors on a single chip only 0.2 inch square. This has led to the "microprocessor," or the central component of a computer. One of the more simple applications of a microprocessor is in display functions. A microprocessor may be permanently programmed to "look" at the cockpit switch settings of a navigation system. The program causes the microprocessor to search out the switch settings at a clock rate, then form impulses to actuate a numerical display. Beyond this simple application, the microprocessor may be permanently programmed to solve an equation.

The microprocessor receives its input from a memory device. The development of inexpensive memory devices has opened up great possibilities in lightweight navigation systems as well as on-board communication and control systems.

Prior to takeoff, a pilot may set in a program of assigned frequencies, after which the system can be tuned to the prearranged frequency with the touch of a button. Memory systems may be applied to any control system that benefits from a prearranged program. For example, after receiving IFR clearance, a pilot may set into memory not only ground control, tower, and assigned en-route communication and navigation frequencies, but a cockpit programming system may also include autopilot setting, etc.

A memory system is an arrangement of devices so that when a given input condition exists, there will be a given output. When compared to human or animal psychology, a more proper word for a digital memory system would be "conditioning," that is, response to a given stimulus. One digital bit may be either logic 0 or logic 1. A memory bit would then respond to a 0, and vice versa with a 1, and this would be the memorized response. By combining a series of bits, an output series would be the response to that stimulus. A very large number of bits can be combined, thus storing a useful amount of information.

The memory system itself is usually a semiconductor matrix in which elements are capable of retaining a condition after a charge has been delivered to it. There will be three modes: *hold, read,* and *erase.* The unit memorizes or stores a digital number or "word" when in the hold mode. When in the read mode, the digital number is retrieved by examining the memory condition with clock pulses. The erase mode clears the memory until another number is stored. Memory units now on the market can survive as many as 10^6 word rewrites, and memory is retained under unpowered conditions for as long as 10 years. More than one word section may be present in a memory IC. The memory may then be addressed to retrieve a word stored in a given section. An IC memory will thus have inputs for mode control (read, rewrite, and erase), along with input (storing or writing mode) and output (read mode). Input and output may be on the same lines.

RHO-THETA NAVIGATION

So far, the assumption has been that IFR flights will be operated from VOR station to VOR station, and up until now IFR charts have shown standard air routes in this manner. This has a limiting effect on airspace, since there is much unused airspace outside of these restricting lanes.

But when DME is employed, the pilot is receiving coordinates in both angle and distance to a known point (the VOR station), and this is sufficient information to place the airplane anywhere within operating distance of any VOR station equipped with DME (VOR-TAC station).

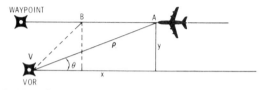

Fig. 5-1. Elementary trigonometry of a "rho-theta" system.

Once distance and bearing to a fixed point are continuously known, it is possible to keep a continuous track of position. Fig. 5-1 shows the elementary trigonometry of a "rho-theta" system. A VOR station, V, furnishes bearing To or From. This results in the angle θ, or theta. The DME furnishes ρ, or rho. A desired course is along the nonradial line AB. Then,

$$y = \rho \sin \theta$$

The course line AB will always be offset by distance y. Since ρ is always available from the DME and y is constant, by continuously

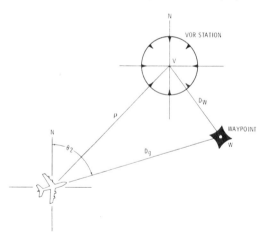

(A) Illustration of the RNAV problem with a known waypoint.

Fig. 5-2. Vector relationships

solving the elements of the triangle, values of θ may be determined and used for a heading to make good course AB.

An aircraft equipped with a system capable of deriving courses

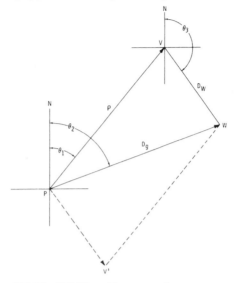

(B) *The RNAV problem treated as vectors.*

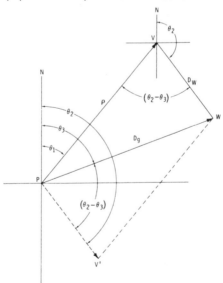

(C) *Quadrantal relationship of vectors in the RNAV problem.*

in the RNAV problem.

offset from a VOR radial will be able to navigate throughout a given area. ATC may then assign courses anywhere in this given area. This process is called *area navigation,* and the on-board system is called *RNAV.* Another advantage of RNAV is that it eliminates the use of marker beacons. When using RNAV, the aircraft can begin its landing approach by intercepting the localizer at a point chosen by RNAV. FAA Advisory Circular 90-45, August 1969, sets up the requirements for area navigation systems for use in the U.S. National Airspace System.

A complete area navigation system should allow the operator to set in the bearing and distance of a waypoint with reference to a given VOR station, and thereafter receive an indication of deviation from the course *to* or *from* the waypoint. If deviation is presented in terms of the distance off of a given ground track *to* or *from* the waypoint, this would satisfy the general requirement for an RNAV system.

Course deviation is independent of the direction in which the aircraft is flying at the moment. The relationship of north in the solution of angular relationships is only incidental. The angular relationship of aircraft, VOR station, and waypoint, together with respective distances, is all that is necessary in solving the RNAV problem (see Fig. 5-2A). The relationship of north is simply established by reference. It is then used as the heading to be flown and held to maintain the chosen track angle which has been selected to reach the waypoint.

The problem can be solved by considering each element as a vector. A vector is a quantity which has both magnitude and direction, and may be represented with a directional quadrant. The sum of two vectors is found by drawing a parallelogram, as shown in Fig. 5-2B. In the parallelogram, PV′ is parallel to VW and PV is parallel to V′W. PV and PV′ are vectors, and PW is therefore the sum. This vector sum represents heading and distance to go to the waypoint. Fig. 5-2C combines the parallelogram with a quadrant with its origin at P, the aircraft position. Angular relationships should be carefully studied.

The navigation problem may be solved by a computer using ac voltage analogs, since a pure sine wave is conveniently represented by a vector. For example,

$$E \sin \omega t = \bar{e}$$

The value $\bar{e}$ is the magnitude of the vector at time t, and is the phase of the voltage which is represented by the direction of the vector. The phase relationship of two sine waves of the same frequency at time t is $\omega_1 t = \theta_1$, and $\omega_2 t = \theta_2$ where ω_1 and ω_2 are expressed in radians of angle. A voltage may be derived using rho and theta, and another using waypoint distance and bearing. The vector sum of the voltage will be a vector voltage representing heading and distance to go.

TYPICAL VOR/DME RNAV SYSTEMS

An RNAV system may be composed by the addition of a small lightweight computer to a VOR/DME installation. By suitably combining equipment, the standard OBS will display the bearing to the waypoint and To/From the waypoint, and the DME indicator may be caused to display the distance to go. It is also possible to display distance and bearing on a separate indicator. In more elaborate systems, several waypoints may be programmed and called up as the flight progresses. Input from an air data computer may also be used to derive true airspeed. This will permit the cautious computation of wind speed and direction, allowing quick choice of better cruising altitude or course.

A typical example of a lightweight RNAV system is the King KN-74, shown in Fig. 5-3. It is designed to operate with King models of VOR and DME receivers, although certain others may be used. Distance and bearing of the waypoint with respect to a chosen VOR station are set in with the respective panel knobs. The Function switch enables each of these five modes:

VOR/DME—The RNAV function is disabled and ordinary navigation to a VOR is used. The DME reads distance to the VORTAC station.

ENROUTE—The DME indicator reads distance to go to waypoint, and the OBS reads the course heading and To/From.

APPROACH—This is essentially the same as the ENROUTE mode, except that the course width displayed on the course width deviation indicator is narrowed from ±1.25 nautical miles.

TEST—This is a self-test of the RNAV computer.

The KN-74 uses ac voltage analog computation. Phase analogs are obtained from the received VOR station. Distance analogs are obtained from the DME output and potentiometer setting for way-

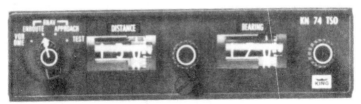

Fig. 5-3. Panel view of the King KN-74 RNAV computer.
(*Courtesy King Radio Corp.*)

point distance. Phase is combined with magnitude by chopping a dc voltage which is analogous to magnitude. The resulting square wave is then filtered to obtain a sine-wave fundamental representing a true vector in phase and magnitude.

Fig. 5-4 is a simplified block diagram showing how the KN-74 solves the navigation triangle. Input to the system is the composite VOR signal and a dc analog voltage representing rho. The 9960-Hz VOR reference signal is derived using a limiter and discriminator. The phase of this signal represents the magnetic north reference, as described in Chapter 4. The VOR radial signal is derived from a low-pass filter with stable phase characteristics. When a VOR station is employed for RNAV, the VOR signal is received differently. Ordinarily, the VOR station would be approached along a fixed radial. In RNAV, however, the aircraft is flying across the radials while moving toward the waypoint. The difference between the 9960-Hz fm reference phase and the received radial phase is now changing. As the VOR station is passed perpendicularly, the rate of angular change is rapid. Moreover, the VOR signal may not be received with equal intensity at all points between radials. The aggregate of this would be to introduce phase instability in the am portion of the composite VOR signal, as well as other effects. Such effects are removed from the received radial by signal processing circuitry. A solid state servo filter is employed which averages phase jitter and smooths phase changes.

Heading to the waypoint is obtained when the OBS resolver shifts the phase of the received radial by the amount $\theta_2 - \theta_3$. The OBS

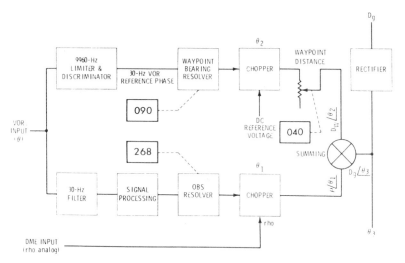

Fig. 5-4. Simplified block diagram of the King KN-74 RNAV computer.

is set to the course heading when a waypoint is selected. Course deviation thereafter may be obtained by measuring the out-of-phase component of the OBS phase-shifted voltage and the D_g vector. This is derived from a phase detector.

When two voltages of different phase are applied to a phase detector, the output will be a dc voltage which represents the out-of-phase component. The dc output will be directly equal to the difference in phase and the magnitude of the ac voltage applied to the phase detector. This is illustrated in Fig. 5-5. Note that the out-of-phase component is always at right angles to the reference vector. This is the principle used in determining course deviation.

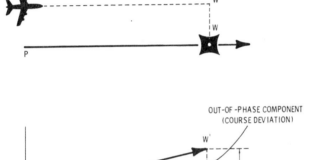

Fig. 5-5. Course deviation phase detector output.

The course deviation needle will deflect in direct proportion to the distance of the aircraft off of the selected course to the waypoint.

Course deviation is a function of both the magnitude of PW and $\Delta\theta$ (Fig. 5-5). Component E_D is perpendicular to E_{REF}. When the course is parallel to PW, E_D is constant. Note that when $\Delta\theta = 90°$, E_D (deviation) is equal to PW, or the distance to the waypoint. Hence, there is no "cone of confusion" as the waypoint is passed, as there is when a VOR station is passed.

The To/From indication is obtained by comparing the OBS phase difference with θ_3 in a phase detector. If the difference is $\pm90°$ or less, a To indication is displayed. When the difference is greater than $\pm90°$, From is displayed.

A more sophisticated approach to RNAV is to use a microprocessor for digital computation. The King KNS-80 Digital Area Navigation System, shown in Fig. 5-6, uses the digital system approach. The microprocessor, VOR, and DME systems are contained on circuit boards mounted in a single panel-mounted unit. An external course deviation indicator is the only external equipment required. Using large scale integration (LSI) techniques, the entire unit weighs only six pounds. A block diagram of the KNS-80 is shown in Fig. 5-7.

The unit is capable of storing up to four waypoints, including the VOR frequency, angle, and distance from VOR to waypoint. The DME tunes automatically to the paired VOR frequency which is set by waypoint designation. Distance to go, time to station or waypoint, and velocity to station or waypoint is displayed.

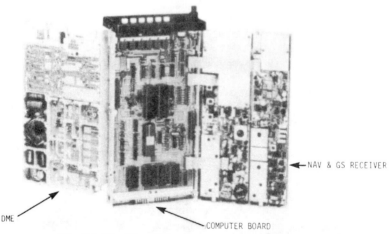

Fig. 5-6. The King KNS-80 Digital Area Navigation System.
(*Courtesy King Radio Corp.*)

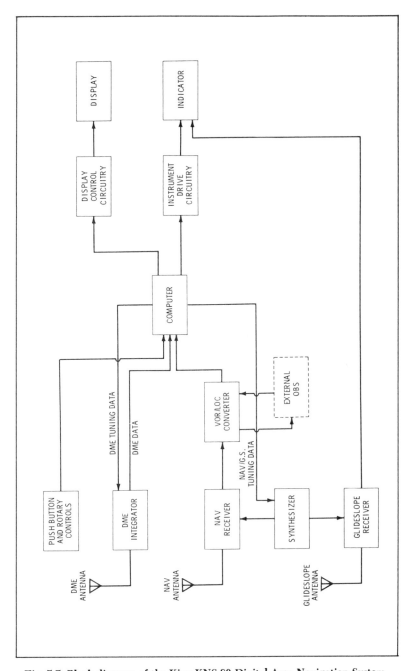

Fig. 5-7. Block diagram of the King KNS-80 Digital Area Navigation System.

GREAT-CIRCLE NAVIGATION

Rho-theta navigation is based on the assumption that the earth is a flat (plane) surface, and dealt with by methods of plane trigonometry. As long as the distance between waypoints is relatively small and speeds are on the order of about 200 knots, the assumption holds. But the earth is, in fact, a great sphere with a radius of approximately 3959 statute miles. The shortest distance between two points on the surface of a sphere is measured over a *great circle*. For this reason, accurate navigation over long distances or higher speeds must be by means of great circles, and carried out in terms of spherical trigonometry. Triangles on the surface of a sphere are formed by the intersection of great circles.

Locations on the earth's surface are based upon coordinates of *latitude* and *longitude*. As will be shown, this bears a direct relationship to great-circle distance between destinations. Because radius is a constant parameter of a sphere, and therefore all points on the surface of a sphere are equidistant from the center, distances on a sphere are a direct function of the angle between two points when measured at the center of the sphere.

One nautical mile is assigned as the amount of distance subtended on the earth's surface by an arc of one minute taken from the center of the earth. The circumference of the earth is, therefore, 60×360, or 21,600 nautical miles. The relationship between ground speed and distance on great-circle paths now becomes greatly simplified. For example, an aircraft flying a true north heading at 300 knots ground speed (a knot is one nautical mile per hour) covers 300 minutes of latitude, or 5 degrees of latitude ($300 \div 60$) per hour. Starting from, say, 40 degrees north latitude on a true north heading, the aircraft would reach the North Pole in 10 hours.

When Charles Lindbergh first planned his historic New York-Paris flight, he visited a public library and, using a string held across a world globe, measured the distance between New York and Paris by calibrating the string and selecting checkpoints. His simple method obviously worked. Nevertheless, a more accurate method must be used for modern high-speed flight. Great-circle distance can be found mathematically by the following equation:

$$(\text{hav}), \text{hav } D = \text{hav}(L_2 - L_1) + \cos L_1 \cos L_2 \text{ hav}(\lambda_1 - \lambda_2)$$

$$(\text{Eq. } 5\text{-}1)$$

where,

(hav), haversine $A = \sin^2 \frac{1}{2}A$,
L_2 is the latitude of destination,
L_1 is the latitude of present position,
λ_2 is the longitude of present position,

λ_1 is the longitude of destination,

D is the distance in terms of an angle (directly convertible to nautical miles).

The haversine is found from a table of haversines. This function is equal to:

$$\text{hav } A = \frac{(s-b)(s-c)}{bc}$$

where,

$$s = \frac{d+b+c}{2} \quad \begin{array}{l}(\text{a, b, and c are the sides of} \\ \text{any plane oblique triangle})\end{array}$$

When flying a great-circle track, the northerly heading is constantly changing. The heading, M_1, at any instant is equal to

$$\sin M_1 = \cos L_2 \sin(\lambda_1 - \lambda_2) \csc D \qquad (\text{Eq. 5-2})$$

Without an on-board computer, the heading must be calculated periodically and incremental changes made to make good a great circle. With a computer, the heading for a true great-circle track may be displayed continuously.

GREAT-CIRCLE VLF NAVIGATION

The earth is now blanketed by powerful vlf signals originating from stations operated by the U.S. Navy and Coast Guard. As pointed out in Chapter 2, vlf ground waves are very stable and the phase of these waves bears close relationship to terrestrial location. When an aircraft is in motion, the change in relative phase angles can be accurately noted and used in determining distance and direction of motion. The advantage of vlf is that it may be used anywhere on the surface of the earth such as in polar regions, and does not depend on line-of-sight to a station. At the present time, 17 stations are available, and each station reaches approximately half the globe with a usable signal.

The U.S. Navy operates nine 1-megawatt vlf stations between 14 kHz and 24 kHz. Although these stations are used for communication with ships and submarines at sea, the phase-stable signals are available for navigation. In addition, eight vlf stations radiating up to 10 kW, known as *Omega* stations, are specifically available for navigation. Fig. 5-8 shows the relationship of the Omega stations. These stations transmit on 10.2, 11.33, and 13.6 kHz. The signals are synchronized into a format so that no two stations are transmitting on the same frequency at the same time. The eight-station format is repeated every 10 seconds in a manner that makes it possible to recognize each station.

By noting the occurrence of transmissions, the composite signals may be interpreted. It is not necessary to receive all stations in order to get a vlf position fix. By comparing the phase from two or more stations, fixes can be made. The technique usually calls for

OMEGA STATION[*]	Segment	GEOGRAPHIC COORDINATES	OMEGA SIGNAL FORMAT							
			Segment Duration							
			10 seconds — 0.2S spacing							
			A	B	C	D	E	F	G	H
			0.9s	1.0s	1.1s	1.2s	1.1s	0.9s	1.2s	1.0s
NORWAY (ALDRA)	A	66 25 15 N / 13 09 10 E	10.2‡	13.6†	11.33†	12.1	12.1	11.05	12.1	12.1
LIBERIA	B	06 18 19.39 N / 10 39 44.21 W	12.0	10.2	13.6	11.33	12.0	12.0	11.05	12.0
HAWAII (HAIKU)	C	21 24 16.9 N / 157 49 52.7 W	11.8	11.8	10.2	13.6	11.33	11.8	11.8	11.05
N. DAKOTA (LEMOURE)	D	46 21 57.2 N / 98 20 08.77 W	11.05	13.1	13.1	10.2	13.6	11.33	13.1	13.1
REUNION ISLAND (INDIAN OCEAN)	E	20 58 26.47 S / 55 17 24.25 E		11.05			10.2	13.6	11.33	
ARGENTINA (GULFO NUEVO)	F	43 03 12.53 S / 65 11 27.69 W			11.05			10.2	13.6	11.33
AUSTRALIA [†]	G	38 29 S / 146 56 E	11.33			11.05			10.2	13.6
JAPAN (TSUSHIMA)	H	34 36 53.26 N / 129 27 12.49 E	13.6	11.33			11.05			10.2

[*] Omega station operational status reports are broadcast on WWV at Fort Collins, Colorado starting at 16 minutes past the hour, and on WWVH at Kekaha, Kauai, Hawaii starting at 47 minutes past the hour. Omega station status reports may also be obtained from the U.S. Coast Guard via recorded telephone message by dialing (202) 245-0298.

[†] At time of publication the G station segment was without transmission. Omega station Australia was projected to become operational in February 1982.

[‡] All frequencies are in kHz.

Fig. 5-8. The Omega Navigation System transmission format.

the use of phase-locked loops to convert the signals from all stations to the same frequency using an accurate clock frequency. No clock is absolute, however, and in time, phase drift and resulting inaccuracy would occur. For this reason, systems are designed to measure only relative phase shift at intervals as the vehicle moves from the point of departure to the next fix. Relative phase shifts from different signals due to motion are analyzed by a computer. The computer interprets phase change and updates at defined periods, usually on the order of 10 seconds.

The Global Navigation Model GNS-500A VLF/Omega Navigation System is an example of an FAA-approved vlf system. This

Fig. 5-9. Control panel of the Global Navigation Model GNS-500A VLF/ Omega navigation system. (*Courtesy Global Navigation, a Unit of Sundstrand Corp.*)

system provides worldwide point-to-point great-circle navigation. Tuning, monitoring, and solving the great-circle equation are completely automatic. Prior to takeoff, the operator sets the internal clock to Greenwich Mean Time (GMT) and date, and sets in the geographical coordinates of the runway. Waypoints may also be set in. The system selects and monitors the appropriate vlf or Omega stations on the ground and during the flight. Fig. 5-9 shows the cockpit control-display unit. A receiver-computer unit is located in an appropriate area, and leads to the external antenna.

Vlf transmissions used by the system are provided by either Very Low Frequency U.S. Naval Communication (Comm) stations or the stations constituting the Omega Navigation Network. The computer automatically selects up to eight stations which will provide optimum navigational accuracies in the geographical area of the flight.

The computer monitors the output of internal digitally phase-locked receivers and continuously compares the incoming signal phase with the phase angle predicted for a geographical position at any given instant.

Phase angles are computed by referencing an internal quartz-crystal oscillator. This oscillator is stabilized and corrected through information derived from the phase tracking process. By comparing actual versus predicted phase angles from a number of stations, the computer determines distance and direction from the starting point. The computer calculates current geographical coordinates, updates the present position display, and stores present position in the memory every 10 seconds. By monitoring changes in latitude and longitude, the computer determines track angle, and ground speed is calculated by measuring the elapsed time of changes. Drift angle becomes available from the heading input. When the system is provided with a true airspeed input, wind direction and velocity are calculated.

Magnetic variation is computed as a function of position coordinates and time. The computer always stores the latest magnetic variation in its memory, thus eliminating the need to set in variation at the start of a flight. This automatic variation computation may be overridden manually if desired.

When an optional rubidium frequency standard is provided, continuous navigation may be achieved when only two stations are being tracked, provided their geometry is appropriate for a solution. When less than two transmitted signals are being tracked, the system automatically operates in a dead-reckoning mode and continues to provide guidance with respect to the selected track. This dead-reckoning function is more accurate when inputs of magnetic heading and true airspeed have been provided from an external source.

The system clock/calendar, in conjunction with an optional standby battery, permits time and date to be kept even when aircraft power is off. The operator may manually update the time and date if necessary. The computer memory contains a program which automatically compensates for most of the diurnal (day/night) shift of each vlf signal. This requires the operator to initially check and update the GMT and date if necessary, then set in present position. The date is used to compute the earth's inclination to the sun. Present position and GMT determine if the aircraft is in daylight or dark and whether pathways between stations and aircraft are being swept by the day/night terminator.

To keep the number of wires to a minimum, all data transmitted between the control unit and receiver unit is serialized and sent in multiplex format.

The system power supply, contained in the receiver unit, incorporates an automatic cutout circuit to protect against undervoltage. A diode protects the system against reverse polarity of the power source.

The system provides present position latitude and longitude on a worldwide basis. Provision is made for the continuous memory storage of the departure point coordinates plus nine waypoints including the destination. Present position is stored in memory updates every 10 seconds. The system computes and displays the great-circle course between selected waypoints and provides a digital readout of both crosstrack distance and distance to go.

The display selector switch, in conjunction with the mode selector switch, is used to call information from the computer or to inform the computer of the type of information to be entered. Ten operating positions of the display selector switch and two positions of the mode selector switch define the data to be viewed in the left and right numerical displays (LND and RND, respectively). Information is displayed according to the combination of settings on the two switches.

Aircraft present position, to the nearest tenth of an arc-minute, is shown on the left (latitude) and right (longitude) numerical displays. The readout can be frozen to facilitate position check by pressing the HOLD key. The North-South-East-West displays will flash to indicate that the displayed position is frozen, but the computer continues to compute position. After the display selector switch is momentarily moved to another position and back to POS, an updated position will be displayed. The present position can be manually updated over a known fix by pressing the HOLD button and inserting the true coordinates.

Waypoint latitude and longitude are shown respectively on the left and right numerical displays to the nearest tenth of an arc-minute. Waypoints 1 through 9, shown in the defined waypoint display window, represent the corresponding waypoint positions that are stored in the computer memory. Waypoint 0 automatically defines present position coordinates whenever a course from present position direct to a defined waypoint is desired. Bearing from the aircraft's present position to the selected TO waypoint is shown in the left numerical display to the nearest degree. Bearing from the selected FR waypoint to the selected TO waypoint, to the nearest degree, is shown in the left numerical display.

Distance to go is shown on the left numerical display to the nearest tenth nautical mile, and is measured between the present position and the selected TO waypoint. Time to go, to the nearest tenth of a minute, derived from the distance from the TO waypoint and current ground speed, is shown on the right numerical display. Estimated time of arrival (ETA), to the nearest tenth of a minute, is shown on the right numerical display when the 0 (zero) key is depressed. If the 0 (zero) key is released, the display will return to estimated time en route (ETE). Since ETA is calculated as the sum of ETE and GMT, DIS, ETE, and ETA information is only available if a FR/TO leg selection has been made. Drift angle is shown on the left numerical display to the nearest degree. The R or L preceding the readout indicates that the track is right or left of aircraft heading.

Wind information can be made available if the following conditions are met:

1. Aircraft heading is automatically set into the computer.
2. True airspeed is set into the computer either automatically or manually by the pilot.
3. Ground speed is in excess of 50 knots.
4. Local magnetic variation has been computed automatically by the system or entered manually.
5. Wind speed is greater than 4 knots.

Wind direction, to the nearest degree with respect to true north, is shown on the left numerical display, and wind speed, to the nearest knot, is shown on the right numerical display.

6

Instrument Landing Systems

To terminate the flight of an airplane, the pilot must reduce speed, descend near the end of the airport runway, and with minimum vertical velocity allow the wheels to touch the runway. Any remaining thrust is then cut off and the airplane is brought to a stop. There are many techniques for landing the various classes of airplanes with their vastly different characteristics. Nevertheless, some common means must be available to guide the pilot to the runway when visibility is poor. This requirement had to be met before the airplane could become an important means of transportation.

Development of instrument landing systems began as long ago as 1929, when General (then Lieutenant) James Doolittle was making the first "blind" landings with a biplane. His methods required considerable agility on the part of the pilot, however, and multi-engine transports were to require more exact radio paths to the landing point. By the early 1930s, crude radio systems for erecting an approach path were in existence in the United States. Shortly thereafter, several Lorentz (German) ILS systems were operating in Europe. By 1942, the present ILS system was developed by the United States, and by the end of the war had been standardized. Marker beacons were first developed in connection with four-course radio ranges, and in addition now have the important function of marking the approach path as part of the standard instrument landing system.

When visibility is minimal, landing is the most critical phase of a flight. There are three classes of landings under instrument flight rules:

Nonprecision approach.
Precision ground controlled approach (PAR).
Instrument landing system approach (ILS).

A nonprecision approach may be made in several ways, as explained elsewhere in the text (these approaches require higher minimum ceilings and forward visibility). The PAR, or ground controlled approach, is unique in that the responsibility for a safe landing is shared by the ground controller as the aircraft is "talked down." As a matter of policy and tradition, shared responsibility, such as in a ground controlled approach, is not favored in commercial and other professional aircraft operations, and only a few airports are PAR equipped. The ILS, therefore, is considered to be the standard means of precision landing. Chapter 1 pointed out the categories of precision landings, based upon visibility.

THE ILS CONCEPT

The instrument landing system consists of a *localizer* transmitting array and a *glide slope* transmitting array. Radiation patterns of these two arrays are shown in Fig. 6-1. The function of the localizer is to guide the airplane laterally on a path along the runway. The glide slope acts to guide the airplane along a path that meets the runway at a shallow angle of approximately 3°. A diagram of the complete ILS is shown in Fig. 1-10 on the foldout at the back of the book.

Both of these radiation systems operate on the same principle. There are two lobes or beams. One lobe is modulated with 150 Hz, the other with 90 Hz. The point where 90-Hz and 150-Hz signals are received with equal intensity always lies upon the same plane in space, and the point where these two planes intersect with the runway is the touchdown point.

The radiation field for localizer and glide path is actually formed by an ingenious method whereby a central antenna array emits a single frequency exactly equal to that specified for the localizer and glide path. The side lobes are formed by emitting frequencies from the side arrays that are within 150 and 90 Hz (the center or carrier frequency).

The combined effect, due to phase relations, is to cause the frequencies to intermodulate within the aircraft receiver, causing equal strength between the 150-Hz and 90-Hz modulation products when the airplane is on the correct path. The choice of 150 Hz and 90 Hz

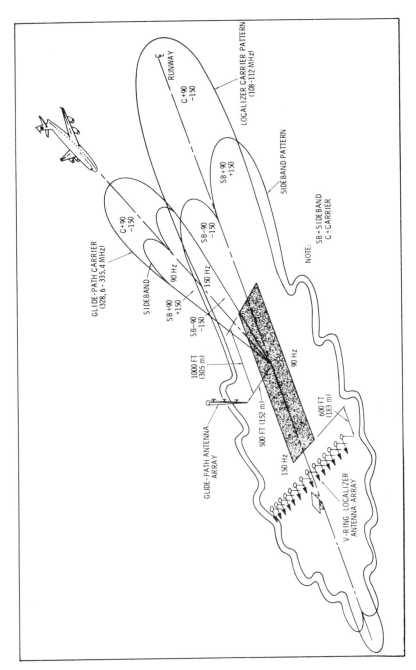

Fig. 6-1. Radiation patterns of localizer and glide-slope arrays.

125

is made so that there will be no undesirable interference between intermodulation products. At the same time, these low audio frequencies are easily handled and do not cause interference with voice modulation. The localizer is also voice modulated without interference. This is done by modulating the basic carrier frequency at a maximum depth of 44%. The localizer may then be used to communicate with the landing aircraft on the same receiver used for the approach guidance.

The localizer ground installation consists of antenna array, transmitter, monitor, and shelter for the transmitter. The antenna array is located on the downwind end of the runway (the opposite end from the aircraft approaching for landing). Sometimes the array is elevated to afford a better signal. The area must be completely free of buildings, trees, power lines, etc.

The glide slope ground installation must also be free of objects that may interfere with the beam. This installation is located on the upwind end of the runway, and consists of transmitter and shelter together with the transmitting mast.

The glide slope installation also includes a monitor that operates the same as the localizer monitor. These monitors constantly receive the localizer and ILS beam, and if either beam fluctuates or the equipment becomes defective, the airport control center receives immediate automatic warning. Site criteria for the ILS system are shown in the previously mentioned Fig. 1-10.

Localizers operate in the vhf range and the glide slope beam operates at uhf frequencies. In order to prevent ILS interference from airports located near or adjacent to each other, different frequencies are assigned to each airport. Larger airports may have more than

Table 6-1. Relationship Between Localizer Needle Deflection and Position of Aircraft

Localizer Needle Deflection	At the Runway Threshold	Distance from Runway Threshold					
		¼ nmi	½ nmi	¾ nmi	1 nmi	2 nmi	3 nmi
¼-Scale Deflection	88 ft	101 ft	114 ft	127 ft	141 ft	194 ft	247 ft
½-Scale Deflection	175 ft	202 ft	228 ft	255 ft	281 ft	387 ft	493 ft
¾-Scale Deflection	263 ft	302 ft	342 ft	382 ft	422 ft	581 ft	740 ft
Full Deflection	350 ft	403 ft	456 ft	509 ft	562 ft	775 ft	987 ft
Localizer Width	700 ft	806 ft	912 ft	1018 ft	1125 ft	1549 ft	1974 ft

To find the **distance of the aircraft from the center of the localizer,** enter table with amount of needle deflection and distance of aircraft from runway threshold. (Note: This table is based on a 4°-wide localizer.)

Table 6-2. Relationship Between Glide Slope Needle Deflection and Position of Aircraft

Glide Slope Needle Deflection	Distance from Runway Touchdown Zone					
	1/4 nmi	1/2 nmi	3/4 nmi	1 nmi	1 1/2 nmi	2 nmi
1/4-Scale Deflection	5 ft	9 ft	14 ft	19 ft	28 ft	37 ft
1/2-Scale Deflection	9 ft	19 ft	28 ft	37 ft	56 ft	74 ft
3/4-Scale Deflection	14 ft	28 ft	42 ft	56 ft	84 ft	111 ft
Full Deflection	19 ft	37 ft	56 ft	74 ft	111 ft	149 ft
Glide Slope Thickness	37 ft	74 ft	111 ft	149 ft	223 ft	297 ft

To find the **distance of the aircraft above or below the glide slope,** enter table with amount of needle deflection and distance of aircraft from runway touchdown zone (not runway threshold).

one ILS-equipped runway, hence, they have more than one ILS frequency. Localizer and glide slope frequencies are paired as shown in Table 6-4.

The third element of the ILS is the marker beacons. All marker beacons operate at a frequency of 75 MHz. Their function is to mark

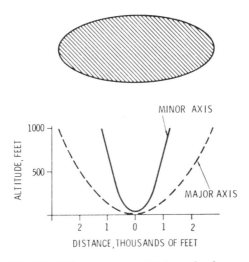

Fig. 6-2. Radiation pattern of ILS marker beacon.

points in space along the localizer and glide slope beams. The location of markers on the ILS layout is shown in Fig. 1-10. Marker beacons used with ILS are low powered. The newer types are pole-mounted solid-state transmitters. Table 6-3 shows the locations of each of the three markers in the ILS. The marker receiver in the aircraft is designed to activate a white, purple, or amber light when the signal from a respective marker is received.

Table 6-3. Marker Beacon Data

Type	Function	Nominal Location	Location Tolerances	Operational Requirement	Modulation	Cockpit Signal
Outer marker	Marks intercept point of glide-slope and minimum-holding altitude	The intercept point	±800 feet (244 meters) lateral and longitudinal	Required for Category I and Category II ILS locations	Two 400-Hz dashes per second	Purple (or blue) light
Middle marker	Marks decision-height point for Category I operations	The decision-height point	±300 feet (91 meters) lateral and ±500 feet (152 meters) longitudinal	Required for Category I and Category II ILS locations	1300 Hz, alternate dots and dashes, repeating 95 times per minute	Amber light
Inner marker (pole-mounted)	Marks decision-height point for Category II operations	See note No. 1	±25 feet (7.5 meters) lateral and longitudinal	Required only for Category II ILS locations	3000 Hz, six dots per second	White light

1. ILS inner marker antenna is located at the point where glide slope elevation is 100 feet (30.5 meters) above the elevation of the highest point in 3000-ft (914-meter) runway touchdown zone.

2. Radar (ASR and PAR) and DME may be used as a substitute for outer marker under certain circumstances.

3. The outer marker may be approximately 4 to 7 miles (6.4-11.3 km) from the ILS landing threshold, while the middle marker may be approximately 3500 feet (1067 meters) from the same threshold.

In addition to the markers used with ILS, there is another class of marker operating at higher power (known as Z or FM) that is used at points along airways. This latter class is no longer in general use. It is modulated with a 3000-Hz tone. A white identifying light appears within the cockpit when the aircraft is over this class of marker.

The marker antenna emits a narrow beam straight up. Fig. 6-2 shows the vertical field pattern of a marker. It should be pointed out that the vertical field pattern of a marker is not, strictly speaking, cone shaped. Diagrams of marker field patterns reveal all points of equal intensity. Thus, at higher altitudes and with a receiver of a given sensitivity, the area over which the marker appears to the pilot actually grows less.

TECHNIQUE OF PRECISION APPROACH

A precision ILS approach may be defined as a landing technique in which the vertical angle of approach to the runway is displayed to the pilot by a moving pointer or other indicator that permits the pilot to adjust his flight path to intersect the runway at the touchdown point.

The IFR pilot must have the "approach plate" or chart of the respective airport ILS procedure in order to make his landing. This chart will show the location of the outer marker, terrain, associated VOR stations, and other facts necessary for landing. The chart will also have a vertical profile showing minimum altitudes.

The key point in the ILS approach is the outer marker. It is from here that descent along the glide path begins. A certain amount of maneuvering is necessary to get to that point. In most cases, approach control will use radar to vector an aircraft to the final inbound course. If there is more than one flight arriving, it will be necessary to "hold" one flight over a point while the other executes a landing. If the flight cannot be vectored, the usual method is to cross the outer marker at an outbound direction. After the marker is passed outbound, a "procedure turn" is executed, as shown in Fig. 6-3. During this turn, altitude is adjusted so that the glide path will be intercepted just over the outer marker. If this is properly done, the approach may be smoothly begun.

The glide path is followed on the instrument needles with the airplane in approach configuration; that is, a given flap setting and airspeed. If the needle appears above center, the pilot must fly up and vice versa. Lateral corrections are made by flying toward the needle. In holding on the glide path, it is important to note that the ILS instrument reveals the position in space, and not aircraft attitude. The pilot controls the aircraft attitude by reference to gyro

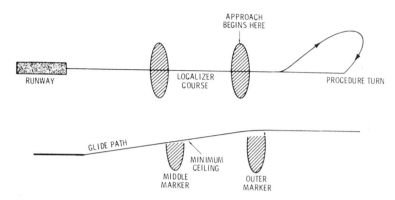

Fig. 6-3. The ILS approach.

signals. This is important, and will be discussed in Chapter 12 in connection with autopilots and flight directors.

The aircraft is maintained on the glide slope until decision height (DH) is reached, at which time the runway must be visible. If not, a missed approach procedure must be executed.

The localizer and glide slope indicators are designed to deflect and show very accurately the positional status of the aircraft. Tables 6-1 and 6-2 show the relationship between instrument deflection and aircraft position. The localizer beam is designed so that it will have a width of 700 feet at the runway threshold at the two extreme deflection positions of the needle. The glide slope needle is three times as sensitive as the localizer needle. Deviation between full deflections of the glide slope needle is 1.4°. When the aircraft is still 2 nautical miles from the runway touchdown zone, a needle deflected half scale indicates that the aircraft is only 74 feet above the glide slope.

Different classes of aircraft have a wide range of landing approach speeds. This has an effect on the technique of controlling landing traffic. The FAA uses categories of approach speed to determine such criteria as landing minimums and obstacle clearance areas for missed approach and circling approach. The approach speed for a given aircraft is set at 1.3 times the stalling speed (V_{so}) at maximum certificated landing weight. The various approach categories are listed on instrument approach charts as follows:

Category A—Speed less than 91 knots.
Category B—Speed 91 knots or more but less than 121 knots.
Category C—Speed 121 knots or more but less than 141 knots.
Category D—Speed 141 knots or more but less than 166 knots.
Category E—Speed 166 knots or more.

MARKER RECEIVERS

The primary requirement of a marker receiver is that it should not convey a false indication by energizing the indicator lights from sources other than the marker-beacon radiation. Field strength is high within the pattern of the beacon, and 2500 microvolts is sufficient receiver sensitivity. A simple transistor circuit for a 75-MHz

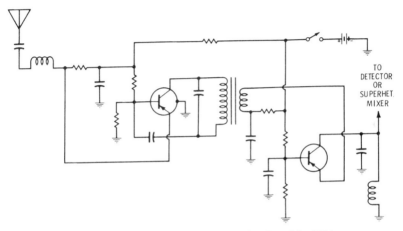

Fig. 6-4. A 75-MHz rf amplifier circuit developed by RCA.

rf amplifier developed by RCA is shown in Fig. 6-4. This circuit may feed the mixer of a superheterodyne, or its output may go directly to a detector. Harmonics from strong local stations (or other phenomena) could cause false indications; but because a marker receiver operates at only one frequency, filters like the one in Fig. 6-5 may be used to increase the selectivity.

The superheterodyne circuit is now almost universally used, sometimes with dual conversion. A typical superheterodyne marker receiver built for airline service will have a 6-dB selectivity of 90

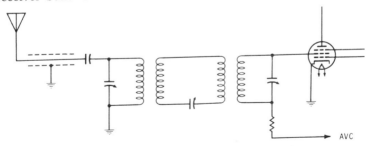

Fig. 6-5. An input filter for 75-MHz marker receivers.

kHz. The local oscillator will be crystal controlled, and the if will be 6 kHz or greater to ensure image rejection. Agc is required with marker receivers to prevent saturation when the airplane is passing directly over the marker station at low altitudes, yet afford sufficient output to the lamp circuits under a range of input levels.

It may be necessary to increase the sensitivity when flying at high altitudes over the airways. Some receivers are therefore provided with a Hi-Low switch, which increases the sensitivity to around 200 microvolts. Sensitivity is adjusted to provide' marker-light operation for a distance of 1000 to 1500 feet on either side of the marker when the airplane is flying 100 feet over the marker on the localizer course. Once adjusted, the marker receiver must be able to maintain constant sensitivity under defined conditions.

The audio detector of the marker receiver is followed by a power amplifier and thereafter by three filters that separate the 400-, 1300-, and 3000-Hz signals, respectively. There are various methods of driving the lights from the output of the filters. One method is directly with an amplifier. Another method is to rectify the signal and feed a relay, which in turn activates the lamp with power from the electrical system of the aircraft.

Audio output of the marker receiver is fed to the cockpit audio system and may be superimposed on the regular voice communication at the pilot's option. This audio, together with the colored lamp, gives positive identification of the marker.

Antennas for aircraft marker receivers may be one of three types, depending on the airspeed capabilities of the aircraft on which each is used. These are wire antennas, metallic structures enclosed in a molded, streamlined dielectric material, or an especially designed portion of the aircraft structure. Fig. 6-6 shows the details of a 75-MHz wire-antenna installation. It is located at least 6 inches below the underside of the fuselage and as far aft as practicable. A coaxial cable such as RG-58/U is used as lead-in, and its outer conductor is

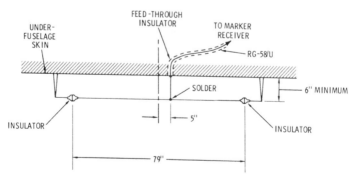

Fig. 6-6. Typical installation of a wire antenna for 75-MHz marker reception.

Fig. 6-7. A streamlined 75-MHz
marker antenna.

grounded at the point where it passes through the skin of the fuselage. The lead-in tap is placed about 5 inches off-center. Fig. 6-7
is an example of a streamlined 75-MHz marker-beacon antenna designed for higher-speed aircraft.

Typical Marker Receivers

The requirements of marker receivers are such that transistors are
appropriate circuit elements. A schematic of the NARCO MBT transistorized marker-beacon receiver is shown in Fig. 6-8. The first stage
(Q1) is an Autodyne in which L3, C7, and crystal X1 cause this stage
to oscillate at 68.75 MHz. The 75-MHz signal from the antenna
appears at J1 and is coupled, through L2, into the base of Q1.
Mixing takes place in Q1, resulting in an if output of 6.25 MHz.
Coil L1 and capacitor C1, in the antenna circuit, are tuned to 68.75
MHz to prevent energy at this frequency from being coupled back
into the Autodyne. Resistors R2 and R3 form a voltage divider that
sets the operating level of Q1. Coil T1 couples the 6.25-MHz if
into if amplifier Q2. Resistor R9 and capacitor C13 form degenerative coupling to neutralize the stage. Resistors R5, R6, R7, and R8
are selected, during production, to establish a standard sensitivity.
Diode CR2 is the detector, and Q3 amplifies the marker signal. In
ordinary agc circuitry, the agc level is determined by the average

133

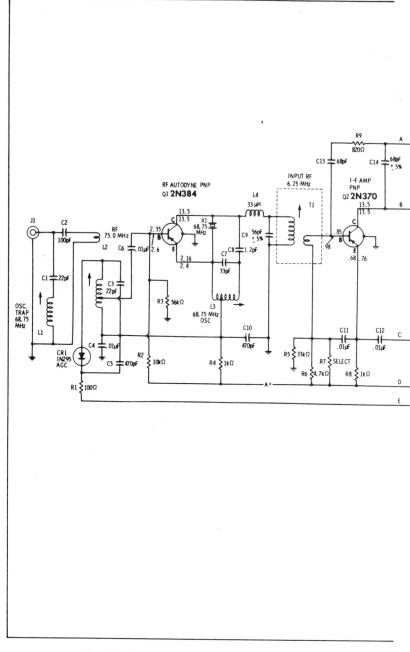

Fig. 6-8. Schematic of the NARCO MBT marker-beacon receiver.

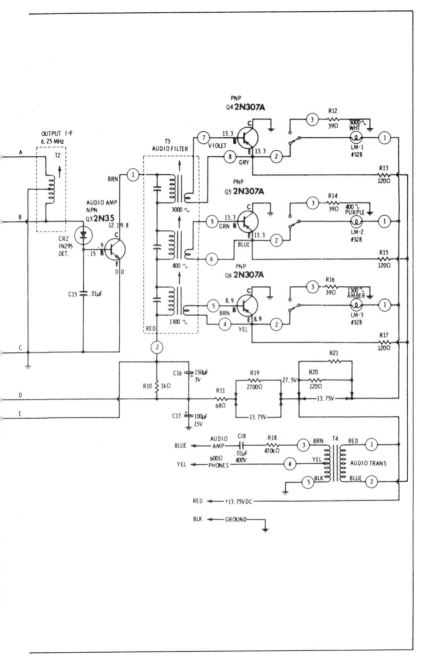

(*Courtesy NARCO Avionics, a Division of NARCO Scientific Industries, Inc.*)

carrier level. In this case, however, agc is developed by Q3, in which the collector current is determined by the demodulated audio. Diode CR1 acts as a clamping diode which establishes a threshold level of agc. As the audio signal increases, so does the current in Q3. The resultant voltage drop across R10 lowers the voltage on the cathode of CR1, causing it to conduct. Conduction of CR1 lowers the voltage on the collector of Q1 because of the voltage drop across R1; hence, the gain of Q1 decreases. Coil T3 is a highly selective filter that selects the corresponding marker-beacon signal. Transistors Q4, Q5, and Q6 are power amplifiers for the respective 6-volt lamps.

Audio energy from Q4, Q5, and Q6 in parallel is applied to the primary of T4. The secondary of T4 has a 600-ohm tap for use with headphones, or a high-impedance tap for coupling to the audio system of the airplane through C18 and R18 (which provide isolation).

The entire unit is packaged in a case measuring only about $3\frac{1}{2}$ by 6 inches and weighing 18 ounces. The three lamps are mounted on the front of the case (see Fig. 6-9), and the entire receiver is mounted behind the instrument panel.

An example of a marker receiver designed for airline use is the RCA AVR-200A shown in block-diagram form in Fig. 6-10. This receiver has a selectivity of 90 kHz at the 6-dB points, and the sensitivity may be set to "High" (200 microvolts) or "Low" (2000 microvolts). This unit weighs only 4 pounds and dissipates a maximum of 10 watts. A quadruple-tuned filter is used at the input to the first converter. A 70.8-MHz local oscillator beats with the signal in the crystal mixer to form a 4.2-MHz if, which is passed through a triple-tuned pi-section filter. A second oscillator-mixer at 4.72 MHz beats the 4.2-MHz signal down to 520 kHz. This signal is then passed through a triple-tuned pi-section filter to three if amplifier stages, followed by a detector. An emitter follower drives the audio amplifier. Agc is obtained by detecting the audio signal and applying it to a dc amplifier. The amplifier controls the agc voltage applied to the first two 520-kHz if stages. Another emitter follower couples the audio signal to three filters that separate the three classes of beacon signals. Each filter output is fed to a monostable multivi-

Fig. 6-9. Panel view of a NARCO MBT marker-beacon receiver. (*Courtesy NARCO Avionics, a Division of NARCO Scientific Industries, Inc.*)

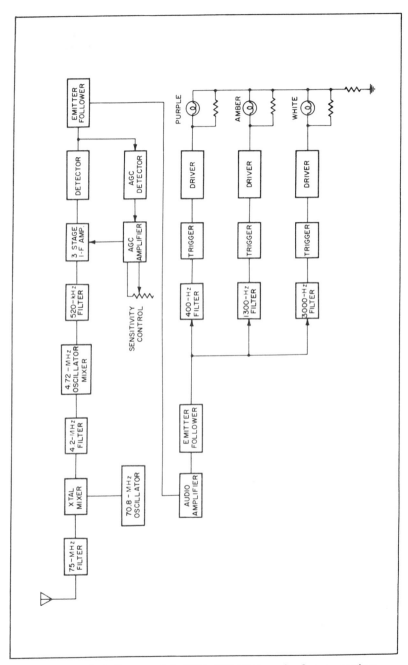

Fig. 6-10. Block diagram of the RCA AVR-200A marker-beacon receiver.

brator trigger circuit. This circuit causes the lamp driver to conduct whenever a signal is present at the trigger input. The lamp driver conducts through the lamp, causing it to light.

Maintenance and Flight Testing

A marker receiver can be bench-checked with a signal generator capable of modulation with an audio oscillator calibrated for 400, 1300, and 3000 Hz. Methods of testing and troubleshooting depend on the type of receiver. Marker receivers as a whole are not difficult to troubleshoot.

Prescribed flight testing of marker receivers begins with a pre-flight ground check and then a flight check, as follows:

1. Turn on the aircraft electrical system, using a power cart if necessary, and allow all electrical and electronic equipment to warm up. Operate all appropriate equipment through ranges which may produce a spurious indication of the 75-MHz marker signal. Check for excess noise in the audio output during these tests.

2. Fly the airplane on the localizer, 1000 feet above the marker beacon. Compute the ground speed over the marker, and make sure the appropriate light remains on for a distance of 2000 to 3000 feet over it. Check for interference that could give false marker indications, by operating the various pieces of equipment as you did in the preflight ground check. Listen for noise in the audio at this time, and note the clarity of the marker signals.

LOCALIZER CIRCUITRY

The localizer course is provided by a vhf signal on 1 of 20 possible frequencies (odd tenths) between 108.1 and 111.9 MHz. A receiver suitably designed for VOR or vhf communication is also suitable for localizer reception. The localizer signal is horizontally polarized, so the same antenna used for VOR can also be used for localizer reception.

As previously pointed out, the localizer appears essentially as two narrow lobes that intersect on the correct course. One lobe is effectively modulated with 90 Hz and the other with 150 Hz. The airplane flies along the path where the 90- and 150-Hz signals are equal in strength.

A circuit for deriving localizer indication from the received signal is shown in Fig. 6-11. The audio output of the receiver is passed to two filters, one of which separates the 90- and the other the 150-Hz components. The filtered signals are then rectified, and the resultant current energizes a galvanometer-type movement forming the right-left steering information for the pilot. The presence of any 90- or 150-Hz signal causes a current flow in R_M and thereby indicates to the pilot that the system is operative. This is accomplished by means of a solenoid that is placed across R_M and is energized whenever a signal is present. A warning flag appears in a window on the instrument when the solenoid is deenergized. The low-frequency localizer signals can be eliminated from the voice channel with little difficulty; the localizer beams are used for communication with the pilot during the approach. Typical bandwidth for the 90-Hz filter is 30 Hz at the 4-dB point and 52 Hz at the 10-dB point. For the 150-Hz filter, typical response is 49 Hz and 84 Hz, respectively, for the same points.

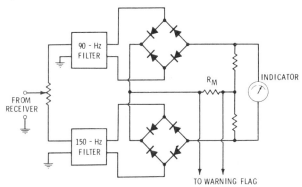

Fig. 6-11. A typical localizer indicator circuit.

Localizer circuitry is ordinarily included in the VOR instrument. When the receiver is tuned to those channels assigned to localizers, their circuitry is automatically substituted for that of the VOR and the same deviation needle used for both. It is possible to make landings with the localizer only, using the altimeter instead of the glide-slope course. The minimum permissible ceilings must be higher, however, because of possible altitude error.

It should be remembered that deflection of the steering needle is proportional to the difference between the 90- and 150-Hz components. This difference, however, will depend on the magnitude of the received signals. Thus, an increase in field strength will deflect the needle more, even though the course of the airplane remains unchanged. To minimize course inaccuracies caused by the foregoing, more elaborate agc circuitry must be employed. One method is to apply agc control voltages to the audio-tone amplification stages. Direct current clamping is sometimes used on the audio agc so that it will become effective when the carrier reaches a certain level.

Fig. 6-12 shows a VOR instrument with localizer circuitry and glide-slope needle. The pilot flies to keep the needles centered. Whenever the VOR receiver is tuned to a localizer frequency, the localizer circuitry functions, and the localizer course deviation is indicated by the VOR needle. Two warning "flags," one for the localizer and one for the glide slope, show "OFF" when no signal is present.

Landing instruments generally exhibit a standard 1000-ohm load. The ILS equipment is usually designed to handle two indicators, and each, therefore, is specified to work into a 500-ohm load.

Fig. 6-12. The NARCO VOA-9 VOR/ILS localizer and glide-slope indicator. (*Courtesy NARCO Avionics, a Division of NARCO Scientific Industries, Inc.*)

GLIDE-SLOPE RECEIVERS

Glide-slope receivers operate between the frequencies of 329.3 and 335 MHz. In accordance with international agreement, glide-slope frequency channels are paired with localizer channels as shown in Table 6-4. In most installations, the glide-slope receiver is automatically channeled to the corresponding glide-slope frequency when the vhf navigation receiver is tuned to a localizer frequency. Superheterodyne receivers are used to receive uhf glide-slope signals, but the rf amplifiers are designed to pass the entire band of frequencies within the glide-slope range. Changing channels is usually done by simply changing the local-oscillator crystal.

Fig. 6-13 shows a block diagram of a typical glide-slope receiver. The rf amplifier is coupled with bandpass filters arranged from sections of coaxial transmission line. The selectivity curve of this rf

Table 6-4. Paired Localizer and Glide-Slope Channels

Channel Number	Localizer Frequency (MHz)	Glide-Slope Frequency (MHz)
1	110.3	335.0
2	109.9	333.8
3	109.5	332.6
4	110.1	334.4
5	109.7	333.2
6	109.3	332.0
7	109.1	331.4
8	110.9	330.8
9	110.7	330.2
10	110.5	329.6
11	108.1	334.7
12	108.3	334.1
13	108.5	329.9
14	108.7	330.5
15	108.9	329.3
16	111.1	331.7
17	111.3	332.3
18	111.5	332.9
19	111.7	333.5
20	111.9	331.1

amplifier is shown in Fig. 6-14. The intermediate frequency is 18.6 MHz. It is possible that the seventh harmonic of the local oscillator could beat with an unwanted signal within the skirt of the rf passband; hence, another section of tuned coaxial line is used at the output of the local-oscillator tripler, which is tuned to suppress the seventh harmonic. The 90- and 150-Hz circuitry for glide-slope reception is identical to that used for localizer reception. Course errors in glide-slope receivers are minimized by using agc in the same manner described for localizer reception.

The Collins 51V3 glide-slope receiver is shown in Fig. 6-15. This receiver meets the following specifications:[*]

Frequency Coverage—329.3 through 335.0 MHz in 20 channels. *Selectivity*—135 kHz at 6 dB; 500 kHz at 60 dB. Image rejection,

---

[*]*Maintenance Manual Glide Slope Receiver*, 51V3, #520-5444-004, 1 February 1959, Collins Radio Co., Cedar Rapids, Iowa.

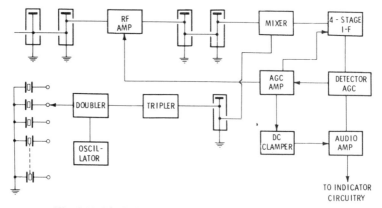

Fig. 6-13. Block diagram of a typical glide-slope receiver.

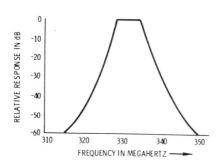

Fig. 6-14. Typical response curve of the rf amplifier in a glide-slope receiver.

Fig. 6-15. The Collins 51V3 glide-slope receiver. (*Courtesy Collins Div., Rockwell International*)

60 dB; adjacent channel, 60 dB; if rejection, 90 dB. All others, 60 dB.

Sensitivity—Deflection sensitivity is defined as the deflection current obtained from a 200-microvolt input signal in which the 90- and 150-Hz modulation differs by a ratio of 2 dB. When the deflection sensitivity is set for 90 microamperes, a 20-microvolt input will produce a 54-microampere deflection current.

Course Stability—±6 microamperes variation from −40°C to +70°C. Variation from any other test condition is not greater than ±5 microamperes.

The receiver employs modular construction, wherein the circuitry is divided into plug-in units. All rf circuitry is in the rf module. A schematic of the Collins 51V3 glide-slope receiver is shown in Fig. 6-16 (see foldout at back of book). A miniature tube, the 5654, is employed for all rf stages. The input stage is broadly tuned to cover the band of glide-slope frequencies. The rf coupling transformers are actually tuned by "distributed constants"; that is, no one part of the circuit is either inductive or capacitive when considered alone. A 5670 dual triode is used as a crystal oscillator and tripler. A crystal is connected between the cathodes of the dual triode, forming a cathode follower that drives a grounded-grid cathode-fed amplifier. Feedback voltage is coupled around from the tank circuit of the grounded-grid amplifier to the grid of the opposite section. The crystal forms a high-Q portion of the oscillator loop, and oscillation is at the frequency to which the crystal is ground. The output of the 5670 is injected into the cathode of the mixer.

Four stages of if at 20.7 MHz are employed, and two stages of audio follow the detector. The first audio stage is a preamplifier with its gain controlled by a portion of the agc voltage from the detector. Also, the cathode self-bias voltage is controlled by R8, the "course softening control." The effective controlling voltage will be the sum of the agc voltage and cathode bias. The circuitry that derives the deviation indication is identical to that described for deriving localizer course indication.

As previously pointed out, glide-slope channels are paired with localizer channels. Aircraft systems are so arranged that when the vhf navigation receiver is tuned to a localizer channel, the glide-slope receiver turns on and a motor-operated switching turret places the correct crystal into the local-oscillator circuit. The band-changing system used in the glide-slope receiver of Fig. 6-16 is shown in Fig. 6-17. Relay 1A7K1 energizes to turn the equipment on when the companion vhf navigation receiver is tuned to a localizer channel.

An open-seeking arrangement is used whereby only 11 conductors plus ground are required to switch into operation any one of

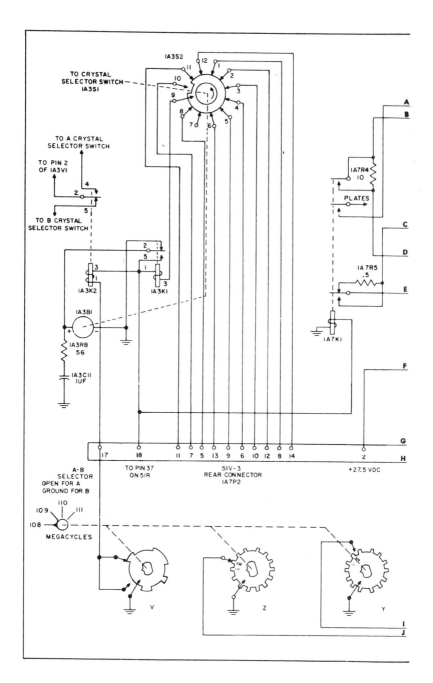

Fig. 6-17. The band-switching system and associated power supply of the Collins

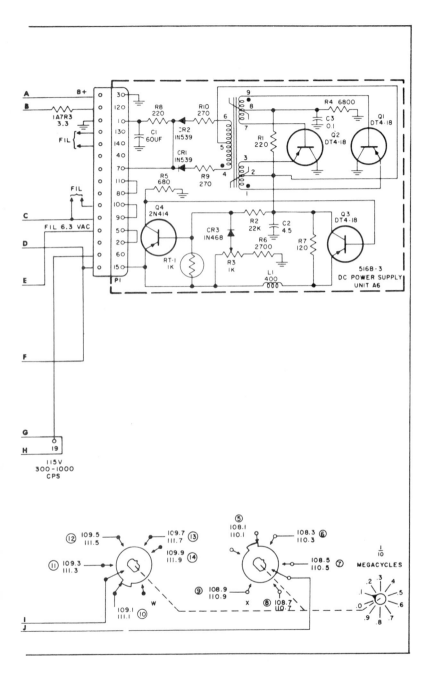

51V3 glide-slope receiver. (*Courtesy Collins Div., Rockwell International*)

20 different glide-slope channels. The band-changing motor, 1A3B1, is energized through the normally open contacts of control relay 1A3K1. Relay 1A3K1 is energized when a new channel is selected. In turn, power is applied to the motor, which rotates switch 1A3S2 and channel-selecting switches 1A3S1 until control relay 1A3K1 is deenergized when the open-seeking switch 1A3S2 aligns with the proper control-wire contact. A short is then placed across the motor field winding which acts as a magnetic brake to stop the motor. Within the vhf control unit on the instrument panel is wafer V of the localizer-channel control switch. When this wafer applies a ground to pin 17 of jack 1A7P2, relay 1A3K2 is energized and selects switch 1A3S1B associated with the 10 B channels. If the ground is removed from relay 1A3K2, it deenergizes and selects 1A3S1A, which is associated with the 10 A channels.

ILS FLIGHT-CHECK PROCEDURE

Prior to the flight, a ground check should be run. With engines running and all equipment on, the system is checked for interference. The disappearance of a warning flag, under any condition when a signal is not present, should be noted and the cause corrected. Before takeoff, the controls should also be checked for proper operation.

A radio-facility chart of the area should be available for the flight. After takeoff, arrangements can be made by radio with a local tower for an ILS check. The flight is continued to a point on the localizer path 25 miles from the touchdown point and 2000 feet above it. At this point, banking the plane up to 10° at all headings and normal pitch angles should not cause the localizer warning flag to appear. A standard instrument landing approach is then made, including maneuvers to check the performance of the deviation needles. During the approach, electrical and electronic equipment should be operated throughout their ranges to check for interference. Propeller modulation will appear as slow oscillation of either or both needles, and changing engine rpm will immediately affect it. If it occurs at landing approach rpm settings, such modulation must be eliminated.

GLIDE-SLOPE ANTENNAS

The first glide-slope antennas were simply horizontal dipoles mounted about 12 inches above the fuselage. Flush-mounted antennas and foil antennas mounted in the cockpit windows have evolved since then, but the most versatile form now seems to be the "U" shape shown in Fig. 6-18. This antenna is horizontally polarized and

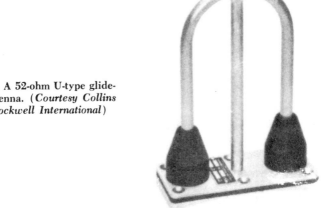

Fig. 6-18. A 52-ohm U-type glide-slope antenna. (*Courtesy Collins Div., Rockwell International*)

has an impedance of 52 ohms into an unbalanced line. It is made of aluminum and weighs only 0.7 pound.

Forward directivity of this antenna is shown in the field pattern of Fig. 6-19. Notice that it offers higher gain forward of the airplane, in the direction of the glide path. Such an antenna can be

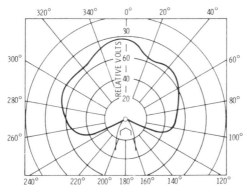

Fig. 6-19. Field pattern of the U-type glide-slope antenna shown in Fig. 6-18.

mounted so it protrudes forward from the nose of an airplane, or included within the plastic nose radome in radar-equipped airplanes. A metallic surface at least 12 inches in radius is desirable for a ground plane. This surface is effectively formed by the nose of the airplane.

147

TESTING AND MAINTAINING
ILS RECEIVING EQUIPMENT

Testing ILS equipment requires a means of simulating actual signals. A signal generator for this purpose is assumed to have an output impedance with a resistive component within 10% of the stated impedance, and a reactive component of not more than 10% of the stated impedance.

The RTCA specifies performance standards and test methods that are generally adhered to. Definitions used in ILS performance tests are as follows:[*]

Standard Localizer Test Signal—An rf carrier, amplitude modulated simultaneously with 90-Hz ±0.3%, and 150-Hz ±3% signals so that the sum of their separate modulation percentages equals 40 ± 2%.

Standard Localizer Centering Signal—A test signal in which the difference in depth of modulation of the 90- and 150-Hz signals is less than 0.002 (0.1 dB).

Difference in Depth of Modulation (DDM)—The percentage modulation depth of the larger signal minus the percentage modulation depth of the smaller signal, divided by 100 (see Table 6-5).

Standard Localizer Deviation Signal—One in which the difference in depth of modulation (DDM) of the 90- and 150-Hz signals is 0.093 ± 0.002 (4 ± 0.1 dB).

Standard Localizer Audio Signal—A localizer test signal to which is added an audio signal modulating the carrier 30%.

Standard Localizer Deflection—"Standard deflection" is 0.6 of the center- to full-scale deflection of the deviation indicator. The receiver shall be adjusted to produce standard deflection when the input signal is a standard localizer deviation signal of 1000 microvolts.

Receiver Sensitivity—The minimum level in microvolts of a standard localizer deviation signal required to produce simultaneously (1) at least 60% deflection of the deviation indicator from the standard deflection, and (2) erratic movement of the deviation indicator due to noise, such movement not to exceed ±5% of standard deflection.

The ARC H14-A signal generator (described in Chapter 4 and shown in Figs. 4-6 and 4-7) can be used to test localizer functioning as well as VOR. Its AMP-LOC circuits generate the vhf localizer

---

[*]Drawn from documents of the *Radio Technical Commission for Aeronautics*, Washington, D.C.

signal modulated with 90 and 150 Hz. In the AMP-LOC mode, the input to the reference-phase channel is removed and the tone generator is replaced by another tone-generator pickup, L103 on tone wheel E101B. This tone wheel is designed to supply a nominal 9960-Hz signal simultaneously frequency-modulated with 90 and 150 Hz. The fm signal is amplified, detected, and fed through filters to remove distortion, then amplified again and applied to the modulation.

Table 6-5. Deviation Indication Versus 90-Hz/150-Hz Tones Percent Modulation°

Localizer			Course Deviation Indicator (CDI)	
90- and 150-Hz % Modulation	DDM	dB	Deflection	Microamperes
20.00 and 20.00	0	0	Centered	0
17.70 and 22.30	0.046	2.007	Half standard	45
15.35 and 24.65	0.093	4.117	Standard	90
12.25 and 27.75	0.155	7.102	Full scale	150
10.00 and 30.00	0.200	9.542	More than full scale	194
0.00 and 40.00	0.400	Infinity	Full one tone; other tone off	387
Standard CDI course deflection = 60% of full scale.				

Glide Slope			Glide Slope Deviation Indicator	
90- and 150-Hz % Modulation	DDM	dB	Deflection	Microamperes
40.00 and 40.00	0	0	Centered	0
37.75 and 42.25	0.045	0.978	Half standard	39
35.45 and 44.55	0.091	1.985	Standard	78
31.25 and 49.75	0.175	3.682	Full scale	150
20.00 and 60.00	0.400	9.542	More than full scale	343
0.00 and 80.00	0.800	Infinity	Full one tone; other tone off	686
Standard G/S indicator deflection = 52% of full scale.				

*Deflection in microamperes for indicators with 150-microamperes full-scale deflection (0.001 DDM is at 2.75 millivolts rms).

The Boonton Type 232-A signal generator is a standard instrument for testing glide-slope receivers. It covers each of the 20 standard glide-slope channels, and a separate crystal may be utilized to generate a signal between 15 and 30 MHz for testing if circuitry. Normally, a 20.7-MHz crystal is provided. "On course," "above course," and "below course" signals can be simulated by varying the ratio of modulation of the 90- and 150-Hz tones with a piston attenuator labeled "DB Tone Ratio Control." For testing course sensitivity and other qualities, the 90/150 and 150/90 audio ratios are usually specified as 2 dB.

As a result of detailed tests, the RTCA found that the calibration of ILS signal generators could be greatly improved by the use of certain prescribed methods. A paper entitled "Calibration Procedures for Signal Generators Used in the Testing of VOR and ILS Receivers" was subsequently issued.* By following the procedures set forth, bearing deviations can be reduced to less than 1%.

Fig. 6-20. The IFR, Inc., NAV-401L Marker/Nav/Comm Test Set.
(*Courtesy IFR, Inc.*)

The ILS equipment of an aircraft is linked together with the VOR receiver through the glide slope channeling system. In turn, the VOR receiver is a part of the communication complement. Hence, these elements should be functionally considered as an overall Comm/Nav system, and may be tested as a single function. The IFR, Inc., NAV-401L Marker/Nav/Comm Test Set (Fig. 6-20) is designed for this purpose. It may be used on the bench or on the ramp, and is powered with rechargeable batteries. The unit measures 11 inches wide by 4 inches high and 15 inches deep and weighs 17 pounds.

The NAV-401L Test Set is designed to meet the functional testing and calibration requirements of Category II ILS systems. It includes a modulated signal generator for MKR, VOR, LOC, G/S, and COMM tests and a variable output attenuator. Maximum output on any band is −7 dBm to −11 dBm.

---

*Obtainable from the Radio Technical Commission for Aeronautics, Washington, D.C.

One crystal frequency is supplied in each band, and variable frequency modes allow all frequencies of all bands to be used. All bands are frequency phase-locked at 25-kHz intervals except G/S XTL mode with LOC switch on. The set includes a counter to measure:

- RF frequency of the signal generator in any band.
- Frequency of any Comm transmitter.
- Any 0 dBm external frequency from 1 MHz to at least 300 MHz.
- Selected VOR bearing as a check on the VOR BRG digital logic functions.

The counter time base is derived from a 10-MHz clock oscillator with an accuracy of ±0.001% from 0°C to 50°C. A power meter is included to measure COMM transmitter power 0–10 or 0–100 W. A peak or average power switch allows COMM XMTR modulation checks, and a rear panel jack permits viewing or listening to the modulation. A modulation monitor accurately indicates modulation levels on any band 0–30% or 0–100%. NAV tones for modulation are digitally derived from a 2.16-MHz crystal oscillator. The 90-Hz and 150-Hz LOC and G/S tones are phase-locked within 0.1°. Relative phases between the 90-Hz and 150-Hz tones can be varied in 5° steps, relative to the 30-Hz reference signal, for a 1° selected step on the VOR BRG selector. LOC and G/S centering may be calibrated from the front panel.

MICROWAVE LANDING SYSTEMS (MLS)

The present ILS system, although in service for many years, suffers from certain disadvantages. It is basically a fixed-beam system. An aircraft must first intercept the fixed beam and then maneuver for a position upon it. The aircraft is then placed in approach flight configuration and flown laterally and horizontally along the beam until the last phases of landing. The path of the beam is set at approximately 3° as previously described, and must be followed regardless of aircraft characteristics. All approaches must be made along this fixed path.

In the course of events, many communities have been built under these long fixed flight paths, and there is an objection to the noise as well as real or imagined danger. This is only one disadvantage of the present system. The long, low landing path also slows traffic flow, and is inappropriate for short-takeoff-and-landing (STOL) aircraft, some conventional light aircraft, and helicopters. The time required for each aircraft to intercept the present fixed beam, overshoot, and proceed from the outer marker to the runway requires other arriving

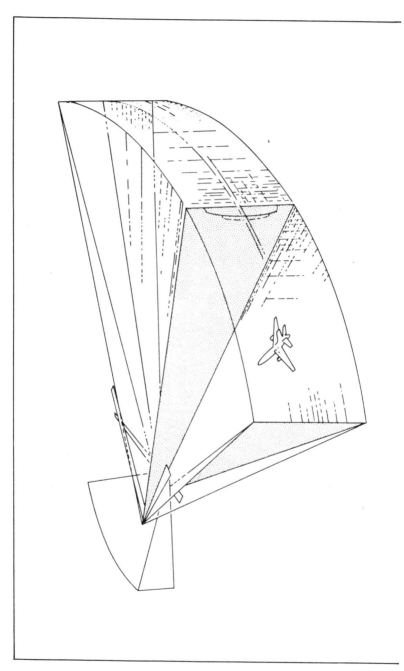

Fig. 6-21. Illustration of the principles of

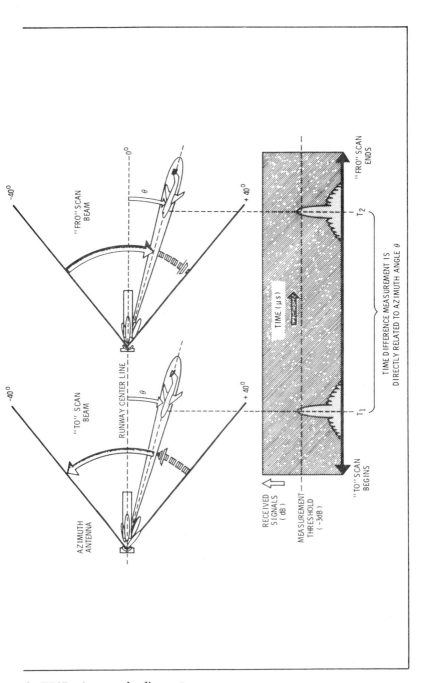

the TRSB microwave landing system.

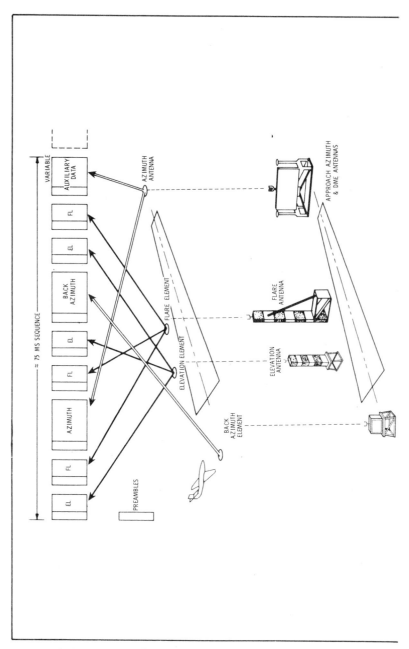

(A) *Location of elements and relationship to multiplex format.*

Fig. 6-22. The TRSB equipment site plan

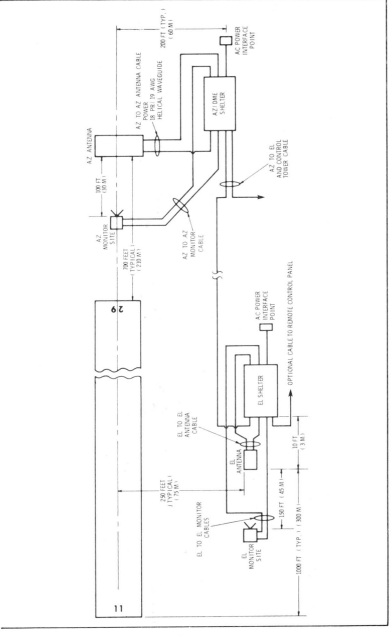

(B) Equipment interconnections.

and signal relations.

aircraft to remain in holding patterns. This results in delay and increased fuel consumption when landing traffic is heavy.

Beyond the traffic requirements just discussed, there is the long-awaited need for a landing system that will allow landing of all aircraft classes in zero visibility (Category III).

Efforts have been made for the past several years to develop a landing system that would not suffer from the disadvantages of the present ILS. In general, the scheme is to use a scanning technique, wherein a microwave beam scans as a function of time. Various nations have engaged in development under recommendations of the International Civil Aviation Organization (ICAO). Two leading candidate systems have emerged for adaptation: the Australian "Interscan" system and the U.S. Time Reference Scanning Beam (TRSB) technique. Both systems are referred to as *microwave landing systems* (MLS). Development in the U.S. is largely funded by the FAA. It is expected that the new TRSB MLS will come into use in the next few years.

PRINCIPLES OF THE TIME REFERENCE SCANNING BEAM (TRSB) SYSTEM

In the TRSB system, the aircraft locates itself in space relative to the runway by receiving microwave "TO" and "FRO" scanning beams which scan relative to time. Fig. 6-21 shows how the TO beam scans with uniform speed starting from one extremity of the coverage sector and moves to the other. It then scans FRO, returning to the starting point. Thus, two pulses are received by the aircraft. The interval between TO and FRO pulses is proportional to the angular position of the aircraft and runway. Another beam scans vertically in the same manner and is used to determine the angular elevation to the runway. The azimuth beam scans at 13.5 Hz and the elevation at 40.5 Hz.

Both beams are at the same frequency; however, the entire system is synchronized so that they are received at different intervals, hence, "multiplexed." Other antenna elements are also incorporated in time intervals over a total period that represents the complete information cycle. Two other elements are "flare" and "missed approach," and there are provisions within the format for other information to be added, such as runway condition, etc. Each segment of information is preceded by a "preamble" which is transmitted by an antenna with broad sector coverage. The system operates in the 5031–5091 MHz "C" band. DME is also included in the system which will be located near the runway touchdown. The DME is conventional, but with refinements for greater accuracy.

Fig. 6-22A shows how the radiation from each respective element intermeshes to form the multiplexed signal at the carrier frequency. Fig. 6-22B shows the interconnection and ground plan of each element. Objects on the ground may reflect energy resulting in false signals; however, the use of the C band permits very narrow beams. Beams (lobes) can be shaped in the nonscan dimension to reduce multipath effects. Tests have been run in the U.S. and Europe in difficult sites which show good accuracy. Automatic landings have been flown with guidance to flare at ± 2 feet down to 8 feet over a 2500-foot touchdown zone.

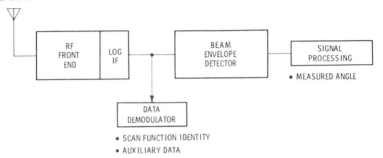

Fig. 6-23. Block diagram of a TRSB receiver.

A block diagram of a TRSB aircraft receiver is shown in Fig. 6-23. The receiver tunes to 200 channels using the double superheterodyne principle. The final if stage includes a logarithmic amplifier with digital demodulator. The signal processor decodes the digital data to determine the function being received. The log video angular guidance signal is digitized and the strongest consistent TO-FRO signals are tracked, then converted to analog if necessary for analog display. The processor is programmed for track acquisition and track validation for immunity from multipath interference.

Several years will be required for the changeover to MLS, depending upon future economic factors. It is reasonable to expect that competition among manufacturers will result in small, inexpensive systems as time goes on.

7

Automatic Direction Finders

Direction finding by radio is as old as radio itself. Before World War I, ships were equipped with direction finders utilizing large loop antennas that were manually rotated from the wheelhouse or an upper deck. Radio direction finders were the first real navigational devices carried by airplanes.

Direction-finding systems make use of the directional characteristics of a loop antenna. For use aboard aircraft, it was necessary to develop small loops, and much had to be learned about their character.

A right/left direction finder was developed in the early 1930s. With this device, a needle would center when the aircraft was pointed to a station, and thereby indicate the direction of flight to "home" on the station. The next step was the development of a system whereby the bearing of a station relative to the aircraft was continuously displayed on a 360° card. The latter is the automatic direction-finder (ADF) system in use today. For many years this system was known as a *radio compass*, but the term has since become less popular in commercial aviation.

By 1937, the old Bureau of Air Commerce made it mandatory for some form of radio direction finder to be employed on all commercial air carriers. Today all long-haul airliners carry two ADF systems. This not only provides a spare in the event one system fails, but also allows instantaneous two-station ADF fixes to be taken. Before the advent of VOR, the ADF was the most versatile navigational aid available. In areas of the world without VOR stations, ADF is still the primary navigational aid. One system employed in other countries consists of three lf transmitter stations that operate on the same frequency and radiate omnidirectional patterns. Each of the three stations transmits in time sequence; therefore, the ADF needle will swing to each station in turn, allowing continuous, automatic, three-station fixes to be made.

When an airplane flies under a low overcast, VOR may be lost because the plane is below the vhf radio horizon. Under such circumstances, low-frequency DF is invaluable.

The ADF can be helpful in many ways. For example, a pilot can determine his distance from any station that can be received on the ADF. This is accomplished in the following manner:

First tune to the frequency of the station, and then turn to the station using the ADF needle. Turn the airplane 90° right or left, and fly a constant magnetic heading. Note the change in ADF bearing. The minutes required to fly to the station will be:

$$\text{Time in minutes} = \frac{60 \times \text{Minutes Between Bearings}}{\text{Degrees of Change in Bearing}}$$

For example, if three minutes were required to cause a bearing change of 15°, the airplane would be $60 \times 3 \div 15 = 12$ minutes from the station.

An ADF system will be most accurate and sensitive when homing on the station. The reason is that the effects of the wing and structure result in less accurate nulls when taken abeam. (The wings should always be level when a bearing is taken abeam.) In general, the stronger the signal, the sharper the null.

If an airplane is proceeding directly toward a station, a crosswind will move the airplane sideways. Hence, if the nose is continuously pointed at the station, the actual ground path will not be a straight line, but will appear as in Fig. 7-1. As the station is neared, changes in heading grow greater and greater; and although the airplane will

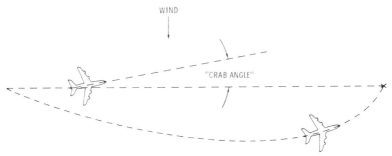

Fig. 7-1. Effect of crosswind on a homing flight path.

eventually reach the station, more time and fuel will be consumed. To prevent a curved flight path, a crosswind correction—equal to the "crab angle"—must be held.

THE LOOP ANTENNA

A loop antenna is in the form of a continuously wound coil and may be square, round, or octagonal. Consider the characteristics of the square form in Fig. 7-2A. Assume that a vertically polarized wavefront is moving from left to right, thus inducing a voltage in AA' and BB'. As pointed out in Chapter 2, the energy in a wavefront is constantly passing back and forth between the electric and magnetic fields. The magnetic lines of force in the moving field will have a different rate of change at AA' than at BB' because of the transit time of the wave. Therefore, the voltage induced in BB' will lag that of AA', and the algebraic sum of the voltages around the loop will not be equal even though the induced voltages are in the same direction. The loop current is the difference between the current through AA' and the current through BB'.

If the wave is moving at an angle θ with respect to the loop (Fig. 7-2B), the lag in voltage BB' and the resultant current will be less than for the previous case. Finally, if the wavefront is passing broadside to the loop, there will be no potential difference and the loop current will be zero. The latter is the null position of the loop. Fig. 7-2C shows the directional pattern of a loop antenna. Rotating the loop to the null position will thus indicate the direction of the station.

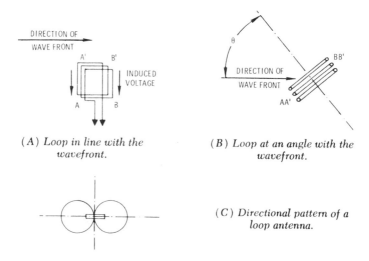

(A) Loop in line with the wavefront.

(B) Loop at an angle with the wavefront.

(C) Directional pattern of a loop antenna.

Fig. 7-2. Characteristics of a square-type loop antenna.

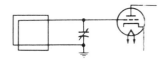

Fig. 7-3. A high-impedance
loop antenna.

The loop is not a very efficient antenna and thus requires large receiver gains. Because it appears inductive, the loop can be tuned to resonance by placing a capacitor across it. This results in a high-impedance output (Fig. 7-3). Low-impedance loops feeding transistor and vacuum-tube stages are shown in Figs. 7-4A and B, respectively. Low-impedance loops for the lf band may have inductances as low as 20 microhenrys. If a loop is not balanced with respect to ground, its directivity will be degenerated. Unbalance occurs when current induced into one side of the loop flows to ground

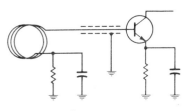

(A) *Feeding a transistor input stage.*

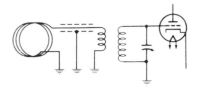

(B) *Feeding a vacuum-tube input stage.*

Fig. 7-4. Low-impedance loop antennas.

without having completed the loop circuit. One method of eliminating the degenerative effects of an unbalanced loop is to employ a shielded loop as shown in Fig. 7-5. The loop is surrounded by a grounded metallic shield that is broken at the center. The wavefront can still penetrate the shield and induce loop currents. Any currents induced in the shield itself will be split into two amounts flowing to ground from either half of the shield. The two shield currents also induce currents in the loop, but the latter currents are equal

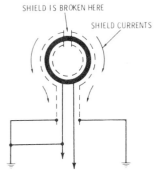

Fig. 7-5. The shielded loop
antenna.

and opposite the ones producing them; therefore, they cancel across the input impedance, thus negating ground currents.

The development of powdered-iron cores has greatly improved loop design and performance. Shielded loops of large diameter were almost universal several years ago; however, the use of ferrite cores has greatly reduced their size. Now loops that can be mounted almost flush with the airplane fuselage are available. The development of accurate goniometers is eliminating the requirement that the loop must rotate.

Precipitation Static

There are two incidental advantages of the shielded loops, one being the obvious rigidity contributed by the shield. The other advantage is the large reduction in precipitation static.

When an airplane is flying through dust, ice crystals, rain, or snow, a condition known as *precipitation static* often develops. Low-frequency receivers accept prohibitively large amounts of this static, and signals are often completely obscured by it. As the speed of the airplane increases, so does precipitation static. At first it was believed that the static was caused by the difference in dielectric charge between the antenna and the particles striking it. A shielded-loop antenna, it was found, decreased precipitation static by as much as 30 dB. Because of this, the low-frequency ADF is designed also to allow reception of weather reports and other information during high precipitation static.

A phenomenon that occurs along with precipitation static is known as *St. Elmo's fire*. Although fairly harmless, this phenomenon appears quite ominous. Propeller tips glow brightly, and sharp points, such as around the airplane windshield, have a bluish glow. When a stroke of lightning occurs nearby, both St. Elmo's fire and precipitation static diminish. This points to the real cause of precipitation static—that is, the charge of several thousand volts accumulated by the airplane in its flight through the precipitation, and its constant discharge into the atmosphere. A nearby stroke of lightning temporarily discharges the atmospheric energy and at the same time the energy accumulated by the airplane.

St. Elmo's fire is simply an extreme form of corona discharge, and it is this discharge that causes precipitation static. Currents from the discharge are in the form of rapid pulsations. Because the corona itself must act as the antenna, the radiated energy from the corona is relatively small, and it is the induced rather than the radiated energy that reaches the receiver. To discharge the accumulated energy without corona, fine wires in series with a resistance of about 100,000 ohms are trailed behind the wingtips or other sections of the airframe. These are the familiar "wicks" that trail from the wingtips of most airliners. Actually, they are ropes impregnated with a conductive material in a manner providing the proper amount of resistance. They are about 15 inches long and are covered with a protective rubber coating.

The Cardioid Pattern

The figure-8 pattern of the ordinary loop is ambiguous; that is, it nulls on two positions 180° apart, and will give the same indication whether pointing toward a station or away from it. By employing a "sense" antenna, however, a cardioid pattern results.

As previously pointed out, the output of a loop is the difference between two voltages induced at opposite sides of the loop. The voltage output of an ordinary open short-wave antenna will simply be its directly induced voltage. This means that the phase of the loop output will always differ by 90° from that of an open antenna. Also, the phase of the loop antenna voltage reverses 180° as it passes

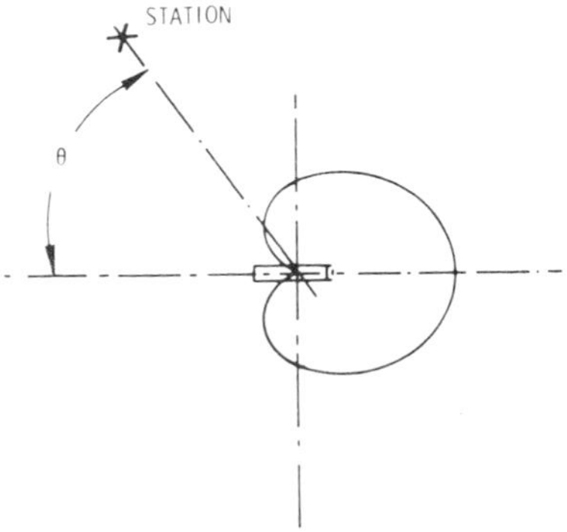

Fig. 7-6. The cardioid field pattern resulting from the combination of loop- and sense-antenna voltages.

through the null. By adding to the loop voltage an additional constant 90° of phase shift, this voltage can be made to add to or oppose that of an open antenna (hereafter to be known as the *sense* antenna). As the loop changes direction, its voltage will vary with respect to the constant sense-antenna voltage, resulting in the cardioid directional pattern shown in Fig. 7-6. Note that only one null position now exists.

The cardioid pattern can be explained by the following simple mathematical relationship:

If θ is the angle of the loop with respect to the station, and E' is the maximum output voltage when the antenna is pointed toward the station, then for any value of θ, the output voltage of the loop will be:

$$E = E' \cos \theta$$

The sense antenna will always receive the same amount, regardless of θ, hence, adding the value of E' to E gives:

$$E = E' + E' \cos \theta$$
$$= E' (1 + \cos \theta)$$

which is the equation of the cardioid.

The foregoing has been simplified by neglecting the fact that E' is actually in the form, E' sin ωt. Therefore, whatever is said about the relationship between loop and sense-antenna voltages, it is important to remember that the cardioid field pattern results only when a constant 90° phase shift has been applied to the voltage of either the loop or sense antenna.

RIGHT/LEFT DIRECTION FINDERS

By switching a cardioid pattern from right to left, it is possible to compare the signals received from the loop and sense antennas and thereby determine (i.e., sense) whether the station lies to the right or left. Consider the circuit of Fig. 7-7. If the ganged switch is opened and closed rapidly (say, with a vibrator device), the pattern will be rapidly switched. The side of the antenna nearer the station will be indicated by a stronger signal. This will be revealed by the steering needle, which will deflect accordingly. The needle will again center when the loop points directly toward the station.

A device employing the mechanical switching described would be somewhat cumbersome. Actual systems utilize tubes or transistors as switching devices. Fig. 7-8 is a simplified schematic of a typical right/left ADF. Switching is accomplished by multivibrator V108. Resistors R131, R132, and capacitors C154, C155 set the frequency at about 48 Hz. Tube V107 acts as a loop switch, reversing the polarity of the loop as the grids are alternately cut off by the signal from V108. Actually this is a modulation process, and the receiver output will now be modulated by the loop signal as it alternates at a 48-Hz rate. Tube V106 amplifies the detected carrier signal that is alternating from each reversal of the loop pattern. Transformers

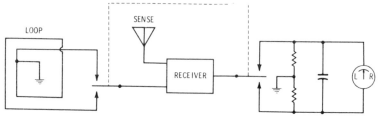

Fig. 7-7. A basic right/left direction finder.

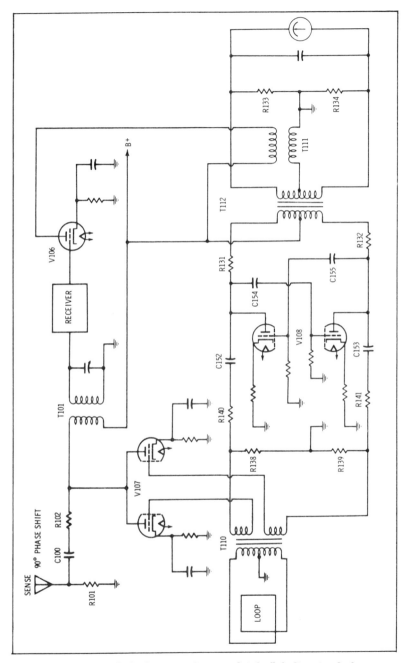

Fig. 7-8. Simplified schematic of a typical right/left direction finder.

T111 and T112 form a phase detector; the polarity of its dc output depends on the phase of the signal from V106 compared with that of V108. The polarity, after comparison, determines the direction in which the needle will deflect; but the amount of deflection depends on the amplitude of the signal from V106 with respect to the constant reference signal from V108.

Examination of a cardioid pattern reveals that as a loop rotates 45° from the position of maximum signal, the output decreases only about 1.6 dB; whereas near the null, only a few degrees of rotation produces a several dB lower output. Because this system zeros on the null, instruments can be made quite sensitive. A system can be adjusted for maximum deflection of the needle with the antenna only 5° off null.

The right/left system will provide two nulls 180° apart; however, this is not as serious as might be expected. The loop is mounted so it will null along the longitudinal axis; and with almost all cockpit instruments, the proper course is established when the airplane is turned toward the needle. With this in mind, the pilot can make the proper correction by observing the following rules:

1. All stations on the right deflect the needle to the right.
2. All stations on the left deflect the needle to the left.
3. All stations dead ahead or dead aft give a needle-center indication.

THE BASIC ADF SYSTEM

Descriptions of ADF operation are often vague and involved. It is well to point out, therefore, that almost all ADF systems are based on a fundamental right/left direction finder with a movable antenna that can be rotated 360°, and a servo arrangement to null the loop.

Fig. 7-9 is a block diagram of a simple ADF system. The reversing relay (M1) is energized in a direction dependent on the polarity of the current flowing through it. In this system, M1 replaces the steering needle of the right/left direction finder. With no current through the relay, there is no contact. Motor M2 has two windings so arranged that current in one winding will rotate the motor in one direction, and current through the other winding will cause opposite rotation. The motor is mechanically connected to the loop and turns it in a direction that is dependent on the position of the relay arm. It can be seen that the loop will rotate in one direction or the other until a null is reached and the relay is neutralized. The mechanical position of the shaft used to rotate the loop will reveal the bearing to the station. This shaft is arranged to position a pointer on a dial calibrated through 360°.

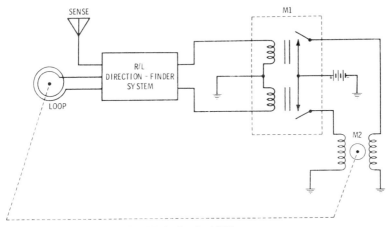

Fig. 7-9. A simple ADF system.

The receiver can be moved to any convenient location in the airplane; it need not be near the loop and driving motor. Likewise, the shaft position can be displayed at a remote location (i.e., the instrument panel) by using synchros. The method just described would be too crude to give accurate results; however, it serves to illustrate the basic principle of ADF.

ADF CIRCUITRY

Instead of the simple switching device shown in Fig. 7-9, the balanced modulator is used to derive the variable-phase directional signal from the loop antenna. Fig. 7-10 shows the balanced-modulator circuit used in almost all ADF systems. The loop amplifier usually has its input tuned, and its output is carefully designed to maintain the 90° phase shift previously discussed. A two-phase reference signal is applied to the grids of the balanced modulator.

The reference signal, usually 45 to 1000 Hz, modulates the carrier signal received on the loop antenna. When a carrier is thus modulated, the output of the balanced modulator will contain only the upper and lower sidebands—the original carrier will be missing. For example, if the carrier is 250 kHz and the reference signal is 100 Hz, the output of the balanced modulator will be two separate frequencies of 250.1 and 249.9 kHz. The center-tapped push-pull transformer (connected to the plates of the balanced modulator) cancels, or "bucks out," the carrier—whereas the modulation products are in phase. The upper- and lower-sideband products are now added to the fixed-phase carrier received on the sense antenna, and a new signal is created and amplified by the receiver. The modula-

167

tion products of the original carrier are still present. The lower-frequency product can be separated from the audio for use as the loop signal.

Circuit action of the balanced modulator is such that the loop signal will be zero when the loop is nulled. On one side of the station the loop signal will be in phase with the reference signal, and on the other side it will be out of phase. The polarity of the phase-detector output indicates the direction in which the loop must turn to reach a null. That is, the circuit can be so arranged that the polarity of the phase-detector output will be, say, positive if the loop is to turn left and negative if it is to turn right.

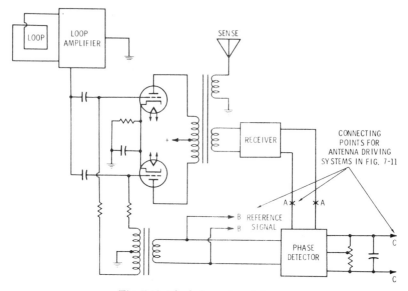

Fig. 7-10. The balanced modulator.

Antenna drive motors are usually of the two-phase induction type with two windings, one to receive the steady reference voltage and the other to receive a signal whose phase determines the direction of rotation. The amplitude of the latter signal also determines the amount of torque developed to rotate the loop. One method of driving the motor is shown in Fig. 7-11A. Here, the reference voltage is amplified to provide the power to drive one phase of the motor, and the signal voltage is amplified for the other phase. In the ARC ADF Type 21A, the reference signal is supplied by a 100-Hz alternator, which provides power to drive one phase of the motor directly.

Another method of driving the loop is shown in Fig. 7-11B. In this system, the dc output of the phase detector is "chopped up" into a

400-Hz signal. This signal, of course, derives its phase and amplitude directly from the output of the phase detector. It is then amplified and applied to one of the motor windings. The other winding is energized from the same source that powers the "chopper." This is somewhat similar to the system used in the Collins ADF Type DF-202, except that the Collins uses electronic means of "chopping."

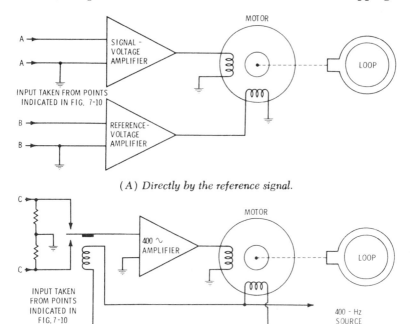

(A) *Directly by the reference signal.*

(B) *By a separate 400-Hz source.*

Fig. 7-11. Loop-antenna driving methods.

A third method of driving a loop is shown in Fig. 7-12 and is employed in the Bendix ADF Type DFA-70. The two thyratrons are biased to fire only when voltage of a certain value is applied to the grids. The plates are energized by a 400-Hz reference voltage. Under these conditions, the tubes will conduct only when the signals applied to both grids and plates are on a positive half-cycle. Transformers T1 and T2 are saturable reactors, and when a tube conducts, the core saturates. Saturation of either core lowers the respective impedance of windings L1 or L2, allowing current to flow through one or the other. Current flowing through L1 will be opposite in phase to that in L2 when referred to the reference voltage. The two thyratrons and associated saturable reactors thus actually form a phase detector for the 46-Hz signal, and current in L1 and L2 is di-

169

rectly related to the phase of the loop signal. Therefore, the motor will turn the loop in one direction or the other until the null is reached, at which time the circuit is balanced and the motor is deenergized.

Since a loop will null at two points, it may appear at first that the system would sometimes read 180° off. This does not happen, however, since at only one null point will the error signal be such

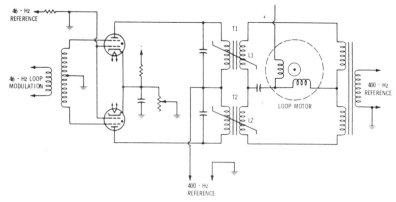

Fig. 7-12. Thyratron loop-drive circuit used in the Bendix DFA-70.

that the loop will turn toward the null. On the opposite null, the servo correction voltage arising from the loop signal will turn the loop away from the null.

GONIOMETER ANTENNA SYSTEMS

Because rotating loop antennas not only are expensive to build and install but also require careful adjustment, they have largely been replaced by goniometer antenna systems (Fig. 7-13). Two fixed loops 90° to each other are employed. Each loop is connected to one of two goniometer windings, also 90° to each other. The two loop antennas receive the vector sum of rf energy and apply it to the goniometer windings. This has the effect of recreating the magnetic field of the wavefront within the shielded enclosure of the goniometer. A rotating winding within the goniometer operates exactly like a loop. It picks up currents, and nulls in relation to the station in the same manner as if an actual wavefront had induced the current.

The advantage of a goniometer system is now apparent. Although the ADF system itself still functions in the same manner as before, the small inertia of the goniometer allows the use of lighter and less-expensive servo loops. The lighter, nonrotating antenna can be

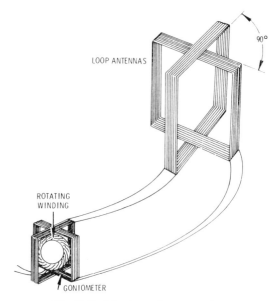

LOOP ANTENNAS

90°

ROTATING
WINDING

GONIOMETER

Fig. 7-13. A basic goniometer system.

attached to the skin of the airplane, thus affording less drag with greater reliability.

Goniometers have been used for almost five decades. Actually, they more recently have been produced under the name *rf resolver,*

Fig. 7-14. The Bendix ADF-T-12B/C ADF system for general aviation.
(*Courtesy Bendix Corp., Avionics Division*)

or simply *resolver*. As far as ADF is concerned, the term *goniometer* seems to have been unearthed from the past to define the use of an rf resolver with two loop antennas separated by 90°. The Bendix ADF-T-12B/ADF system for general aviation is shown in Fig. 7-14. This unit uses a fixed loop with a resolver operating on the goniometer principle.

FREQUENCY SYNTHESIS IN ADF RECEIVERS

In Chapter 3, frequency synthesis using a number of frequency-related crystals was discussed. Another method, made possible by digital ICs is now used on ADF receivers. This method is known as phase-locked-loop (PLL) frequency synthesis. This process is used for supplying a local oscillator in the ADF receiver.

Only one crystal is used, which acts as a clock signal that is compared to the phase of a voltage-controlled oscillator (vco). An error voltage is developed that is used to lock the phase of the vco to the clock signal. The versatility of the system comes from the fact that the clock signal can be arranged so that the vco operates over a range of frequencies, each of which is accurately held within the tolerance of the one basic crystal.

This method of frequency synthesis is coming into wide use in any application requiring a range of closely held frequencies. A feature of PLL frequency synthesis is that the system can be arranged so that frequencies may be chosen by digit in decimal increments. This is a great advantage in avionics, since the pilot may quickly set in an exact frequency merely by glancing at a digital counter or display. Incidentally, although PLL frequency synthesis is covered here in connection with ADF, the method may also be used in vhf Comm/Nav equipment.

A block diagram of a basic PLL system is shown in Fig. 7-15. The phase detector is a relatively simple circuit, contained in an IC

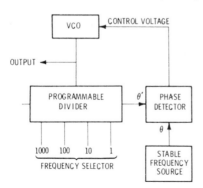

Fig. 7-15. Block diagram of basic phase-locked-loop frequency synthesizer.

package, that compares the phase of two inputs that are of the same frequency and produces a voltage when the two inputs differ in phase. The magnitude of the voltage is a function of the phase difference. The polarity depends upon whether the reference signal leads or lags the other. It will be remembered that:

$$\theta = 2\pi \text{ft}$$

where,

θ is phase,
π is 3.14,
f is frequency in hertz,
t is any instant in time.

Thus, when $\theta = \theta'$ there will be no voltage from the phase detector. With a PLL system, however, the vco settles down at a point at which the phase difference produces a voltage that holds the phase, hence, frequency, constant. The frequency at which the vco settles down depends upon the setting of the programmable divider. Its output frequency will be counted down to reach that equal to the stable frequency source. The block diagram is simplified and does not show additional circuitry necessary in the vco. A practical vco can only operate well over a range of frequencies that may be less than that desired. It is therefore necessary to introduce an adjustment into the vco, such as an offset voltage, that will be made as the frequency selector changes range.

The mass market rising in the Citizens Radio Service has resulted in the development of inexpensive ICs containing both phase detector and vco. It appears that PLL frequency synthesis will be the accepted method for future designs.

STANDARD ADF SYSTEMS

The airlines have standardized ADF systems, to simplify interchangeability and also to make it easier for pilots to become familiar with controls. Some systems are more refined than others because each manufacturer maintains a certain amount of initiative, basing his product on his engineering and price philosophy.

In general, ADF systems have three modes of operation, which can be selected by a switch as follows:

1. *ADF*—In this mode the system functions as a true ADF, with the bearing of a station automatically and continuously indicated on a suitable instrument.
2. *Loop*—Here the pilot manually positions the loop on the null. (The servo system is disabled and the loop drive motor is con-

trolled manually.) This mode is necessary for finding bearings on weak stations. Also, when weather reports or other information must be received during precipitation static or thunderstorms, the loop may be turned to the position of maximum signal. The sense antenna is automatically disconnected in this mode; consequently, there will be two loop nulls.

3. *Antenna*—Only the sense antenna is connected. In this mode, the ADF can be used as an lf/mf receiver only. To enable orientation on A/N range stations, the agc is automatically disconnected and a manual rf gain control is used.

The ADF receivers must represent the ultimate in performance. Because the lower-frequency bands are noisy, such receivers must have excellent selectivity and signal-plus-noise to noise ratios. Stations sometimes can be identified only by their frequency, in which case very close tracking and calibration limits are required. The use of a beat-frequency oscillator (bfo) increases sensitivity to the point where good bearings can be obtained from weak stations.

Bearing indication is provided to the pilot by an instrument with a circular dial calibrated to 360°. By means of synchros, a needle rotates with the loop antenna. The compass-heading card of an instrument known as a *radio magnetic indicator* (Fig. 7-16) is rotated by the magnetic heading provided from a gyrostabilized compass. The large needle displays the ADF heading, while a servo-operated VOR system positions the small needle on a VOR radial. The radio-magnetic indicator thus provides complete heading information on a single instrument and therefore is a great aid under instrument flight conditions.

The Genave SIGMA/1500 is an example of an ADF employing PLL frequency synthesis with a goniometer antenna system (Fig. 7-17). The receiver may be digitally tuned from 200 kHz to 1699

Fig. 7-16. Radio magnetic indicator provides complete heading information on a single instrument (see text). (*Courtesy Collins Div., Rockwell International*)

Fig. 7-17. The Genave SIGMA/1500 ADF uses PLL frequency synthesis to cover all 1-kHz increments from 200 kHz to 1699 kHz. (*Courtesy General Aviation Electronics, Inc.*)

kHz in 1-kHz steps, resulting in 1499 distinct channels. It uses a 141.5-kHz if with single conversion. Performance of this unit in the ADF mode is as follows:

> *Bearing Accuracy*—±3° with 70 microvolts per meter field strength.
> *Image Rejection*—80 dB @ 200 kHz and 400 kHz, 70 dB @ 800 kHz, and 55 dB @ 1600 kHz.
> *Selectivity*—2 kHz min at 6 dB down.
> *Sensitivity*—Not more than 100 microvolts per meter for 6 dB signal-plus-noise to noise ratio.

This performance specification makes the unit capable of meeting all requirements for a light ADF system for use in IFR flight.

A block diagram of the system is shown in Fig. 7-18. The unit is fabricated on two printed-circuit boards—the receiver board and the synthesizer/servo board. Servo antenna positioning with a balanced modulator as used in this unit has been previously described. Electronic bandswitching is employed in all of the tuned rf circuitry. This eliminates mechanical switches and enhances reliability. Tuned circuits are switched in parallel electronically and varactors are used to tune the input circuitry as the frequency synthesizer is set up for a given frequency. The vco output is applied to the mixer at 141.5 kHz above the received carrier frequency. Three stages of if provide about 90 dB of gain. When the bfo function is selected, a 1-kHz beat frequency is applied to the last if stage.

FLIGHT CALIBRATION OF ADF

Loops are subject to quadrantal error; that is, the loop bearing will deviate from the true direction of the station by a differing

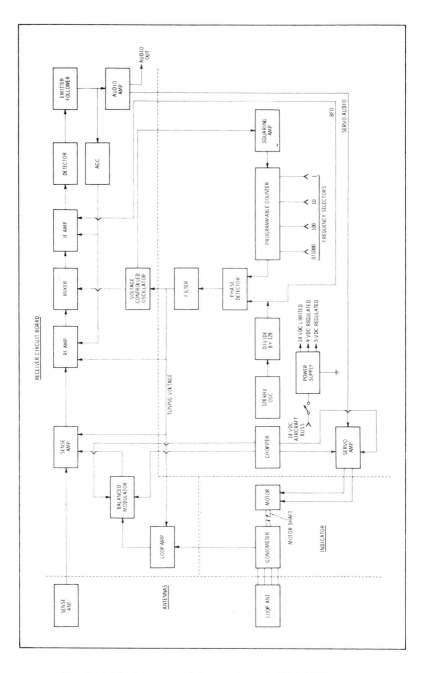

Fig. 7-18. Block diagram of Genave SIGMA/1500 ADF system.

amount around the compass. This is due to the metallic elements of the airplane, which cause the incoming wavefront to be distorted in the immediate vicinity of the airplane. The error is usually removed by placing ferrite "correctors" around the loop, or by otherwise varying the magnetic circuit in such a way that the variations in field strength balance out.

Flight tests are necessary for determining quadrantal error. After the ADF system has been installed and ground-checked, preparations are made for a flight check. Flights should not be made in early morning or late afternoon, because of night-effect errors. The air should be smooth, with winds as light as possible. The flight test should be made away from other air traffic, and at a point where a good ground reference can be established. A fairly strong station at least 50 miles away should be available, and its bearing from the ground reference point established.

After takeoff, the airplane is first flown directly toward some nearby transmitting station having a vertical antenna tower that is easily visible. The flight should be made with or against the wind to preclude crab angle. The axis of the airplane must be directly in line with the transmitter tower. In this position, the ADF indicator should read zero. If it does not, the amount of error should be noted.

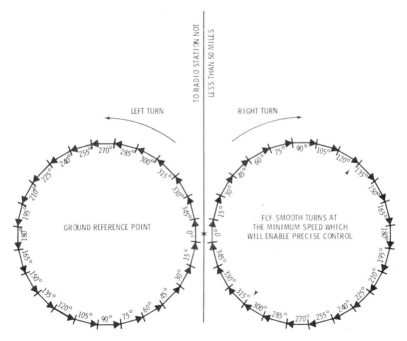

Fig. 7-19. Flight procedure for measuring quadrantal error.

The aircraft now flies to the previously selected point, and the selected distant station is tuned in on the ADF. Using the directional gyro, two 360° turns are flown as shown in Fig. 7-19. The heading indicated on the directional gyro is compared with the ADF heading at each 15° around the circle. Turns should be as smooth as possible, and at the completion of each circle the gyro should be checked for precession and be reset if necessary.

The readings are tabulated and the loop is corrected according to the instructions of the individual manufacturer.

TESTING AND MAINTAINING ADF

Troubleshooting ADF is somewhat more difficult than for VOR. However, if the basic principles previously described are understood and applied to a study of the manufacturer's literature for a specific system, tests and repairs can be made with a minimum loss of time. If the cause of ADF failure cannot be located in the aircraft wiring and interconnections, it can likely be isolated to the control head, loop, receiver, or power supply. When a system fails to null on a bearing in the "ADF" mode, the servo components should be checked by rotating the antenna to a null in the "Loop" mode. If the antenna appears to rotate properly but lacks a sharp null, the trouble is probably a defect in the receiver or sense-antenna circuit. If a bearing can be taken on all but one band, the trouble is likely to be in the rf portion of the receiver. A defect in the sense-antenna circuitry will be revealed by low sensitivity and a higher noise level when operating in the "Ant." mode.

No response from the indicator pointer may mean failure of the synchros associated with the indicator. Failure of the loop servo can be caused by a defective drive motor or gear train, or the reference signal may be absent. Normally, with no signal present the loop will drift slowly; however, a continuous rotation of more than once per minute may indicate a defect in the balanced modulator and its associated components, or a maladjustment.

When the performance of an ordinary receiver is to be measured, a station signal can be simulated by connecting a signal generator (and an impedance-matching network when necessary) directly to the receiver antenna terminals. With ADF, however, the loop must be provided with an actual wave of accurately known field strength in order to establish the true performance. One method is to employ a standard screen room in which a signal is radiated under carefully controlled conditions. Another method of obtaining a controlled rf field is to employ a special simulator such as the Collins 477V-1 loop simulator or the Carter Engineering Service Model CES116A signal simulator.

The ADF testing begins with the receiver. First, function tests (of the various modes, agc voltage, etc.) are made. These are followed by checks of alignment, calibration, tracking, sensitivity, and selectivity. The standard receiver load is 300 ohms with an input signal 30% modulated with 1000 Hz. Tests are next run in each of the three modes, using an actual controlled rf field or loop simulator to measure signal-plus-noise to noise ratios. Dynamic tests are made to determine servo response to a change in bearing. Each manufacturer recommends the method of testing and specifies limits for each test.

8

Distance-Measuring Equipment

Two of the more recent devices to come into use on the airways of the nation are radar transponders and distance-measuring equipment (DME). Although completely different in function, they have technical features in common. Both are pulse devices, and both operate in the same frequency range. These devices are comparatively expensive, but their use is dictated by our more crowded airways and by the higher speeds of modern aircraft. At present, radar transponders and DMEs are mandatory only for high-flying jet airplanes, both military and commercial; all turboprop and pressurized airplanes; and all aircraft with a gross takeoff weight of more than 12,000 pounds.

To use DME, the pilot turns on the equipment and tunes to the station. (In systems where DME is automatically tuned, it is only necessary to tune the VOR receiver.) Each DME station can be identified by its coded audio tone, which is transmitted at 30-second intervals. The DME system first goes into the "search" mode. After about 30 seconds, the system should "lock on" and the indicator will now display the distance from the aircraft to the station. Sometimes a signal may be temporarily interrupted by the maneuvering of the aircraft. In such a case, a "memory" circuit holds the system at the last reading for 10 seconds. If contact with the station is not resumed within this time, the search cycle begins again. The "search" mode is characterized by complete excursions of the distance indicator.

Fig. 8-1. Many DME systems provide an indication of distance on an instrument such as the one shown here.

On the ARINC standard DME system, distances are indicated in nautical miles by a cockpit instrument like the one shown in Fig. 8-1. Distance is measured in a straight line between the airplane and ground station ("slant distance"). It is not the actual ground distance, which is:

$$\text{Ground Distance} = \sqrt{S_L{}^2 - A^2}$$

where,

S_L is the slant distance,
A is the altitude above the station.

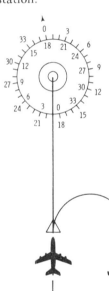

Fig. 8-2. One type of DME holding pattern.

When an airplane is at least one mile from the station for each mile of altitude, the difference between slant distance and ground distance is negligible. Slant distances are measured to an accuracy of ±0.1 nautical mile from 0 to 5 miles, and ±0.2 nautical mile from 5 to 197 miles (the maximum distance which can be measured).

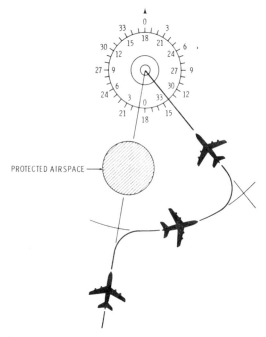

Fig. 8-3. Using DME to avoid an area assigned to other traffic.

The DME provides much versatility in flight operations. For example, DME makes holding patterns possible at any point along a course. In Fig. 8-2 the pilot is instructed to hold at a point 25 nautical miles from the omni station on a given radial. While flying the radial, a procedure turn is entered as soon as the distance indicator reads 25 miles from the station. At the completion of the turn, the airplane is flown outbound the desired distance, and then turned back to the radial. The DME also makes orbital paths possible. By flying to hold a constant reading on the DME, the pilot may orbit an airplane at a fixed radius from the station. With DME, it is also possible to avoid a given air space, as in Fig. 8-3. Here, the pilot is instructed to enter an orbital path 30 miles from a given VORTAC facility, and to proceed along this path to a new radial. The pilot begins a standard rate turn at a point calculated to place him 30 miles from the station on the orbital path and then, using DME, flies this orbital

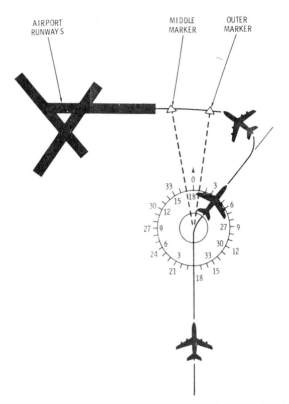

Fig. 8-4. Using DME as a means of maneuvering the aircraft to intersect the localizer on final approach.

path to the new radial. In certain cases, DME may also be used to place the airplane on final approach (Fig. 8-4). The airplane is flown to the station on the designated radial; then it is turned outbound, on another radial, to an orbital path that intersects the localizer.

PRINCIPLE OF DME

Radar provided the first means of measuring distance by radio. It is obvious that the pulse techniques developed in conjunction with radar systems could be put to work for the sole purpose of measuring distance with an accuracy suitable for aerial navigation. After some consideration, the International Civil Aviation Organization (ICAO), in 1949, recommended the development of distance-measuring equipment that would operate in the band between 960 and 1215 MHz.

For several years after World War II, the military services sponsored development of a new short-range navigation system known as *TACAN* (*TAC*tical *A*ir *N*avigation). TACAN uses a pulse system and gives highly accurate bearing and range information from a ground station. It closely paralleled the distance-measuring systems proposed by ICAO, but bearing information was already available to civil aircraft through the VOR stations. When TACAN was proposed for civil use, civil operators objected strongly to the complexity and expense of pulse systems, but agreed that the distance-measurement feature should be retained. Thus, commercial DME systems now use only the distance-measuring portion of TACAN; bearing information is obtained from VOR. A ground station equipped with both VOR and TACAN is called a *VORTAC* station. Military aircraft obtain bearing and distance through the TACAN system, but private and commercial planes utilize the VOR for bearing and TACAN for measuring distance. A station with VOR and only the distance-measurement portion of TACAN is known as *VOR/DME*.

In the DME system, the airborne unit transmits a 2-pulse group to the ground station. After a predetermined delay, the ground station retransmits a pulse group. The pulse travels to the station and returns at a velocity of 3×10^8 meters per second, thus the transit time of impulses is a direct function of distance. A basic DME system is shown in the block diagram of Fig. 8-5. Pulses are sent at one frequency and received at a different frequency, using the same antenna. The airborne station originates two 3.5-microsecond pulses at a rate of about 150 pairs per second. Each pair is received by the ground station, where the pulses are detected, amplified by a video amplifier, and then delayed for an accurately calibrated period. A modulator is thereafter triggered and a pulse group from the ground station transmitted in return. Since many airplanes are using the ground station, the equipment must be capable of selecting only those pulses that are respective replies. This selection is accomplished by the search-and-track circuit in the airborne equipment. It receives a signal from the modulator and, in effect, examines all signals received to determine which have a regular time relation with respect to the transmitted signals. The modulator is designed to pulse at an irregular rate, thus precluding the possibility that its pulses may accidentally fall into synchronism with those from another interrogator. When the search circuit determines which received pulses are due to its own interrogations, the tracking unit locks on them. At the same time, the pulse rate is greatly slowed, which reduces the load at the ground station. Indicating circuits measure the transit time of the pulse groups and thus provide an indication of distance.

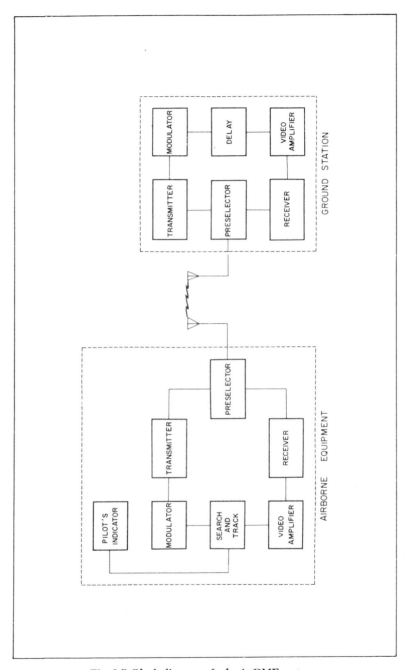

Fig. 8-5. Block diagram of a basic DME system.

Table 8-1. Frequency Pairing Plan for TACAN Channels and Associated VOR Frequencies

Short-Range Terminal (TVOR—ILS) Facilities

TACAN Channel Number	VOR—ILS (Even)(Odd) Frequency (MHz)	TACAN Channel Number	VOR—ILS (Even)(Odd) Frequency (MHz)
17	108.0	37	110.0
18	108.1	38	110.1
19	108.2	39	110.2
20	108.3	40	110.3
21	108.4	41	110.4
22	108.5	42	110.5
23	108.6	43	110.6
24	108.7	44	110.7
25	108.8	45	110.8
26	108.9	46	110.9
27	109.0	47	111.0
28	109.1	48	111.1
29	109.2	49	111.2
30	109.3	50	111.3
31	109.4	51	111.4
32	109.5	52	111.5
33	109.6	53	111.6
34	109.7	54	111.7
35	109.8	55	111.8
36	109.9	56	111.9

Long-Range, En-Route Low-(LVOR) and High-(HVOR) Altitude Facilities

TACAN Channel Number	VOR Frequency (MHz)	TACAN Channel Number	VOR Frequency (MHz)	TACAN Channel Number	VOR Frequency (MHz)
57	112.0	87	114.0	107	116.0
58	112.1	88	114.1	108	116.1
59	112.2	89	114.2	109	116.2
70	112.3	90	114.3	110	116.3
71	112.4	91	114.4	111	116.4
72	112.5	92	114.5	112	116.5
73	112.6	93	114.6	113	116.6
74	112.7	94	114.7	114	116.7
75	112.8	95	114.8	115	116.8
76	112.9	96	114.9	116	116.9
77	113.0	97	115.0	117	117.0
78	113.1	98	115.1	118	117.1
79	113.2	99	115.2	119	117.2
80	113.3	100	115.3	120	117.3
81	113.4	101	115.4	121	117.4
82	113.5	102	115.5	122	117.5
83	113.6	103	115.6	123	117.6
84	113.7	104	115.7	124	117.7
85	113.8	105	115.8	125	117.8
86	113.9	106	115.9	126	117.9

The DME interrogator in the airplane transmits between 962 and 1024 MHz and between 1151 and 1213 MHz; it receives on any frequency from 1025 to 1150 MHz. For commercial and general aviation, there are 100 transmitting channels, each having a designated companion receiving frequency. Transmitting and receiving frequencies are given a channel number that is paired with a VOR channel so that DME can be tuned automatically when the VOR frequency is selected (in the same manner that glide slope is paired with localizer frequencies). Table 8-1 gives the TACAN channel number and its associated VOR receiving frequency. Channels 17 through 56 are paired with low-powered TVOR (terminal VOR) stations (108.0 to 111.9 MHz), the two being classified as "short-range" channels. The DME unit is designed to search only to a preset range between 25 and 70 nautical miles when set to one of these channels.

DME CIRCUITRY

The advent of the integrated circuit has meant the development of several electronic approaches for obtaining an indication of distance in DME systems. The DME makes use of pulse circuitry. Before discussing the DME circuitry itself, certain fundamental pulse techniques will be reviewed.

Pulses can be shaped by utilizing the charge and discharge times of capacitors. When in series with a resistor, a capacitor requires a finite time to charge (Fig. 8-6A) or discharge (Fig. 8-6B). Both actions are in accordance with a logarithmic curve. The greater the capacitance and series resistance, the longer the charge time. The first portion of the curve is essentially straight. Usually the

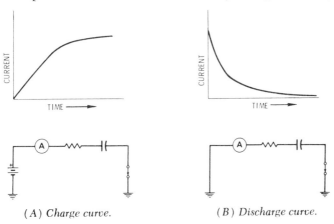

(A) Charge curve.   (B) Discharge curve.

Fig. 8-6. The charge and discharge curves of a capacitor.

first 63% is considered linear. Mathematically, the time required for a capacitor in series with a resistor to reach 63% of its full charge or discharge is equal to the product of the capacitance and its series resistance. This is known as the *time constant*.

If, instead of a switch, a pulse train is provided, we have the same effect as if a switch were continuously opened and closed.

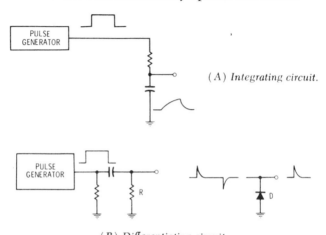

(A) *Integrating circuit.*

(B) *Differentiating circuit.*

Fig. 8-7. Two types of pulse circuitry.

The series would now appear as in Fig. 8-7. Fig. 8-7B, the *differentiating circuit,* can be used to obtain sharply defined pulses. The sharp leading edge of the output pulse coincides with the leading edge of the input pulse. Replacing resistor R with diode D will eliminate the negative differentiating pulses, resulting in a positive pulse train. Reversing diode polarity has the opposite effect. The positive pulses are eliminated, resulting in a train of negative pulses. A sine wave may be turned into a square wave merely by driving a transistor into saturation, as shown in Fig. 8-8A. The square wave can then be converted into trigger pulses as in Fig. 8-8B.

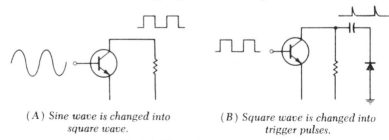

(A) *Sine wave is changed into square wave.*

(B) *Square wave is changed into trigger pulses.*

Fig. 8-8. Wave changing circuits.

A multivibrator is a regenerative circuit that can be designed to operate in one of three states:

Free-running.
Monostable.
Bistable.

A free-running multivibrator is an oscillator. It produces a uniform series of pulses, and may be synchronized to run together with some other pulsating source. Its free-running frequency depends upon circuit values. The monostable multivibrator has one stable state. If triggered with a pulse, the multivibrator will pulse over to another state, but will return to its stable state and remain until again triggered. A bistable multivibrator will trigger to one or the other of two states each time it is triggered. In the description of a hypothetical DME system which follows, a free-running multivibrator and a monostable multivibrator will be included.

A multivibrator operates because it is capable of amplifying its input and applying its output back to the input, hence it is basically an amplifier. The IC operational amplifier is now often used as a monostable multivibrator as shown in Fig. 8-9.

The inverting input (−) terminal is biased at an adjustable level. The noninverting input (+) has a positive input, and the output is therefore saturated near the supply level. The capacitor is normally charged to positive on the right-hand side and negative on the left. When a positive input trigger is coupled to the input terminal, the output falls rapidly negative, pushing the noninverting input terminal down. This keeps the output negative until C1 has completely discharged through R1 and R2. The discharge process starts as soon as the output goes negative. Eventually, C1 charges posi-

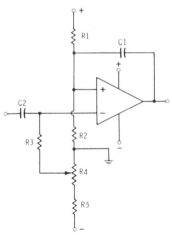

Fig. 8-9. This IC operational amplifier is used as a variable pulse width multivibrator.

tively on the left-hand side and negatively on the right. As soon as the voltage on the left-hand side rises above the voltage at the inverting terminal, the noninverting terminal again has a positive input. The output now rapidly rises to the stable state. Since the bias voltage at the inverting terminal is adjustable, the time during which the output remains negative is controlled, thus making pulse width adjustable.

The fundamental circuitry of the DME is the search/track system shown in the block diagram of Fig. 8-10. The 8088-Hz clock is the basic timing unit. At 8088 Hz, the period of one cycle is 123.7 microseconds, the time required for a radio impulse to travel 20 nautical miles, or 10 nautical miles and return. The 8088-Hz timing signal takes two paths, one leading to the resolver phase shifter and the other to a pulse former. The latter shapes a short pulse for each cycle of the clock, and feeds these pulses to one input of AND logic module U1. When a pulse from the 30- and 150-Hz multivibrator coincides with a timing pulse, AND logic U1 will produce a pulse. This pulse first passes through a coder and then triggers the modulator, causing a two-pulse group to be transmitted. At the same time, the pulse from U1 triggers the variable pulse width multivibrator. The multivibrator output appears on one input of AND logic U2. Meanwhile, the resolver output has been passed to another pulse former, and its output now appears on the other input to U2. The shaft of the phase shifter is geared at a ratio of 10:1 with the multivibrator timing potentiometer. The variable pulse width has the range required for a 200-mile round trip of a radio impulse. Thus, the amount by which the common shaft has rotated will determine the time between the pulses at the outputs of logic modules U1 and U2.

When the pulses from the ground station are properly spaced, the decoder produces one pulse into the error detector. The error detector is also receiving pulses from the output of logic module U2. The timing shaft will be correctly positioned when the ground-station and latter pulses coincide. The error detector, servo amplifier, and servo motor drive the timing shaft to the point where ground-station returns coincide with the locally generated timing pulses. The position of the timing shaft will indicate the distance. When the servo is operative, the search-motor shaft is locked and stationary. Any servo-motor movement will be repeated in the output of the differential. When the station is first tuned in, however, the servo loop is stationary and the search motor causes the timing shaft to go through complete excursions until pulses appear at the error detector. At this point, the servo loop takes over.

The reason for transmitting two-pulse groups rather than a single pulse is that this lessens the chance of an airborne station being affected by replies to interrogations from other airborne stations

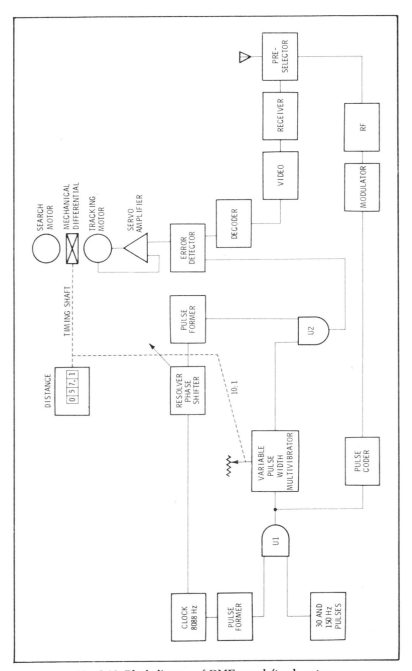

Fig. 8-10. Block diagram of DME search/track system.

191

when several are interrogating the same ground station. The two-pulse method (called *pulse multiplexing*) also allows closer channel spacing. Stations respond to only the pairs of pulses that are spaced at the proper intervals. As the number of interrogating stations increase, however, so does the probability that two pulses arriving from two different stations will have the proper relation to trigger the decoder. Pulse pairs are separated by 12 microseconds, and decoders operate between 10 and 14 microseconds. It should be remembered that the ground station must reply continuously to all stations. As the number of stations increases, more average power is demanded from the ground station. During the searching mode, the airborne station has a pulse rate of about 150 pulse pairs per second. Once the tracking mode is established, the pulse rate is cut to the lowest value consistent with proper tracking. In the example given, this was 30 pulse groups per second, a value set by the 30- and 150-Hz multivibrator. Thus, lowering the pulse rate of the airborne unit lowers the power dissipation of the ground station.

RF CIRCUITRY

Fig. 8-11 shows the arrangement of rf circuits in DME. The channels for transmission and reception are always separated by 63 MHz, and the selected if is also 63 MHz. Therefore, a single frequency synthesizer serves for both the transmitter and for the local oscillator in the receiver. The frequency synthesizer consists of high- and low-frequency crystal-controlled oscillators. This output is mixed to derive a sum frequency which is multiplied 24 times. The low-frequency oscillator uses 10 crystals between 9.86290 and 10.23800 MHz. There are 13 high-frequency crystals from 32.8037 to 37.8037 MHz. The single pulse from the timing oscillator of the search/track system fires the blocking oscillator, producing a pulse that is fed to the modulator. The modulator generates a high-amplitude, positive-going square wave which serves as the plate voltage for the power amplifier.

At the same time the original timing pulse is fed to the blocking oscillator, it is also fed to the delay line. The pulse that emerges from the delay line 12 microseconds later also fires the blocking oscillator, causing the same chain of events as the pulse discussed previously. The resultant output from the power amplifier is two pulses separated by 12 microseconds. Pulse radiation is at the same frequency as the input signal to the power amplifier. The latter stage is actually a lighthouse tube operating in a resonant cavity (sometimes more than one tube is used in parallel, to increase the power). Another method that results in greater power and stability is to use a buffer, driver, and final power stage, each designed as a

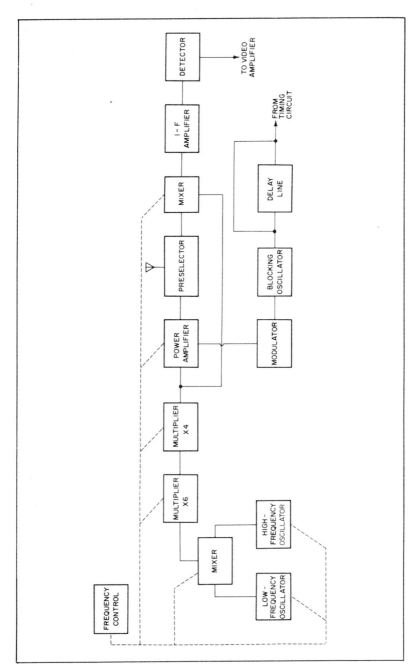

Fig. 8-11. Block diagram showing arrangement of rf circuits in DME.

193

cavity feeding the other. Resonant cavities are highly stable, with unloaded Qs as high as 2500.

The preselector prevents transmitter energy from entering the receiver. It usually accomplishes this by means of one or more gas-filled tubes known as *TR* (*Transmit/Receive*) types. A burst of rf energy from the transmitter ionizes these tubes, and their conducting gases change the electrical appearance of the circuit to the rf energy in such a way that energy is prevented from reaching the receiver.

The if amplifiers in DME usually have a bandwidth of 6 MHz, and their agc requirements differ from those of ordinary receivers. In DME, it is not the average carrier that determines the agc voltage, but the level of only those pulses that result from decoding.

TYPICAL LIGHT DME

NARCO UDI-2 distance-measuring equipment was designed specifically for the general aviation fleet. It provides 100 channels, with

ANTENNA

PANEL UNIT

RANGE-POWER UNIT

Fig. 8-12. The NARCO UDI-2 DME is designed specifically for general aviation. (*Courtesy NARCO Avionics, a Division of NARCO Scientific Industries, Inc.*)

ranges up to 100 miles at an accuracy of ±0.5 mile or ±3%, whichever is greater. The unit weighs less than 16 pounds complete, consumes approximately 117 watts, and operates on 14- or 28-volt dc systems. The UDI-2 consists of three units, as shown in Fig. 8-12. The pilot controls the equipment from the panel unit. The dial is calibrated in VOR channels. The UDI-2 is ordinarily independent of the VOR receiver. To select a DME channel, the pilot sets the corresponding VOR frequency on the UDI-2 dial.

In operation, the distance indicator searches automatically through its range every 20 to 30 seconds, and locks on when the correct return pulses are contacted. Memory is provided for 8 to 12 seconds, after which the search is resumed. Audio power of up to 50 milliwatts is available for identification of the ground station if necessary.

The equipment is installed as shown in Fig. 8-13. The Range-Power unit contains the power supply, search/track circuitry, and receiver if amplifier. It may be provided with a shock mount, if desired, and located at some more convenient remote place. The DME panel unit is strapped behind the instrument panel and contains the rf circuitry. Having the rf circuitry within easy reach means that the dial can turn the tuning shafts directly, thereby eliminating expensive automatic remote-positioning devices. The distance indicator may be removed from the unit and mounted separately in any standard $3\frac{1}{8}$-inch instrument hole on the panel.

The antenna used with the NARCO UDI-2 DME is encased in fiber glass and epoxy, and measures only $3\frac{1}{2}$ by $3\frac{1}{4}$ inches on the outside. The antenna is mounted on the bottom of the fuselage, and the skin of the aircraft acts as a ground plane.

The King DKM-700 DME was also designed specifically for general aviation. It provides 100 channels with an accuracy of ± 0.1 nautical mile or $\pm 0.2\%$ of range, whichever is greater. The unit has an acquisition sensitivity of -80 dBm with a lock-on in four out of five searches. The unit itself weighs 11.8 pounds, operating at 13.75 $\pm 20\%$ volts dc, 5.5 amperes or 27.5 $\pm 20\%$ volts dc, 2.75 amperes while in track mode (note that in the track mode a DME unit is radiating power). Distance is indicated on a panel-mounted KDI-570 Range/Speed/Time-to-Station indicator shown in Fig. 8-14. The KDI-570, weighing about 2 pounds, provides a way of measuring radial velocity relative to the VORTAC station by determining the rate of change in DME range with time. Speed is read from 0 to 600 knots in 1-knot increments. Another feature is the Time To/From Station indication which indicates intervals of 1 minute up to 60 minutes.

A block diagram of the KDI-570 is shown in Fig. 8-15. Distance readout is derived from circuits incoming from the KDM-700. Velocity integration is provided with an integrator circuit consisting of Q301 and Q304. Output of the velocity integrator is a voltage analog from 0 to -6 volts dc which represents the radial velocity of the aircraft relative to the VORTAC station. The velocity voltage analog is converted to mechanical readout by means of the 400-Hz servo loop. Velocity may be changed to "Time-to-Station" by the position of S602. At the same time that distance is being integrated, it appears

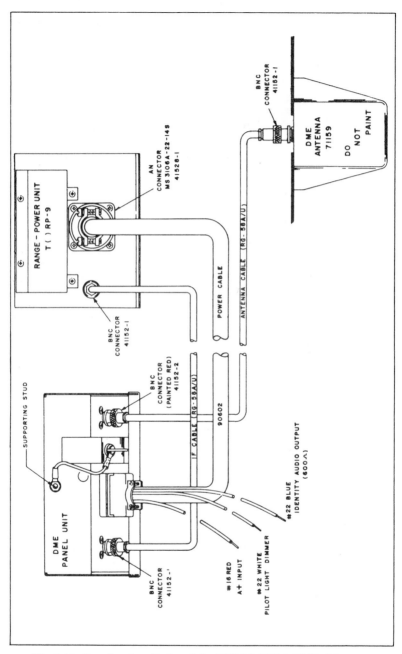

Fig. 8-13. Complete installation of the NARCO UDI-2 DME. (*Courtesy
NARCO Avionics, a Division of NARCO Scientific Industries, Inc.*)

as a decimal readout that is actuated by respective digital-to-analog matrix.

During the initial warm-up period after the unit is turned on, a red and white warning flag will be present in the upper left-hand digit of the KDI-570. The warning flag will also be present when the DME is inoperative. During an inoperative period, the ground speed or Time-to-Station numerals drift slowly. When the unit locks on after the search period, the warning flag disappears and correct range is indicated. The VORTAC station may be identified by listening to its audio identification.

Fig. 8-14. The KDI-570 Range/ Speed/Time-to-Station indicator used with the King KDM-700 DME. (*Courtesy King Radio Corp.*)

About 1½ minutes after lock-on is required for the integrator to compute ground speed to within 10%. After a 3-minute integrating period, true ground speed will be within 1%. While channeling to a new frequency when tracking is in progress, the unit will display zeros in the top row of digits. The flag will not come into view until the tuning motor has transferred to the new channel (about 4 seconds). After a maximum of 6 seconds of search for the more distant stations, lock-on should occur and the flag will disappear.

TESTING DME

The following operating characteristics are usually checked when testing distance-measuring equipment:

Receiver sensitivity.
Transmitter power.
Decoder sensitivity.
Identification.
Distance and velocity accuracy.
Search, ranging, and memory.
Agc performance.

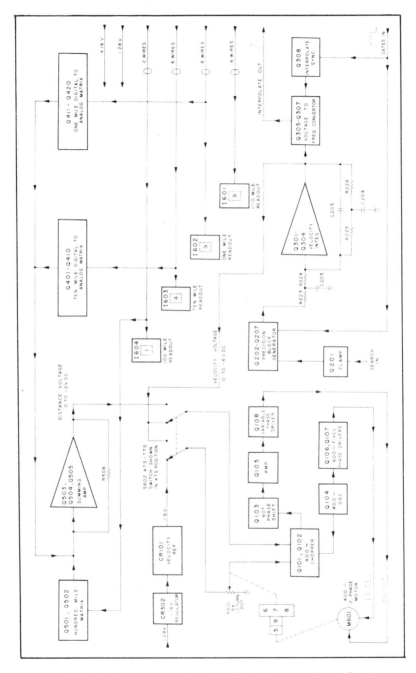

Fig. 8-15. Block diagram of the King KDI-570. (*Courtesy King Radio Corp.*)

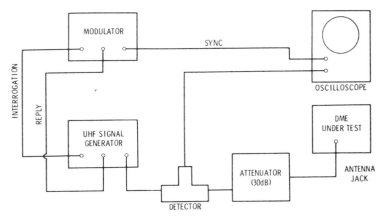

Fig. 8-16. Typical DME test setup.

Test equipment for DME usually consists of a special uhf signal generator and modulator connected as shown in Fig. 8-16. The uhf generator provides the rf signal, which has been changed into pulse groups by the modulator. Together, these two instruments actually simulate the ground station and reply to the interrogations of the DME. One input of the oscilloscope is connected to a detector, which is inserted into the output of the DME so that both responses can be monitored. A known amount of attenuation is also inserted into the DME output to reduce the transmitter output to a conveniently measurable level.

The Tel-Instrument Electronics Corporation Type T-16A DME Pulse Generator shown in Fig. 8-17 is designed to produce reply pulses which simulate the DME portion of the VORTAC ground station. The T-16A is designed to be used with such signal generators

Fig. 8-17. The TIC Type T-16A DME Pulse Generator is a solid-state instrument that provides laboratory-quality impulses for DME checkout. (*Courtesy Tel-Instrument Electronics Corp.*)

Fig. 8-18. The TIC Type T-15B ATC/DME Signal Generator provides a variable carrier for the modulation supplied by the T-16A DME Pulse Generator. (*Courtesy Tel-Instrument Electronics Corp.*)

as the Hewlett-Packard 612A and the Hewlett-Packard 8925A DME/ATC Test Set, or preferably, it may be used with the Tel-Instrument T-15B ATC/DME Signal Generator shown in Fig. 8-18.

Output of the T-16A consists of video pulse trains which modulate the signal generators previously mentioned. The unit will test DME response to simulated distances up to 310 nautical miles and ground-speed indication up to 600 knots. This latter feature is important when testing newer equipment that includes ground-speed indicators. Output pulses are 0-10 volts into 50 ohms, with the coded pulse-pair spacing adjustable from 3 to 39 microseconds.

The IFR, Inc., Model ATC-1200Y3 Transponder/DME Test Set (Fig. 8-19) combines the ability to test DME and transponders

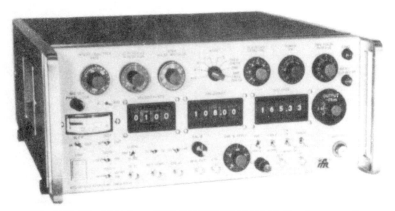

Fig. 8-19. The IFR, Inc., ATC-1200Y3 XPDR/DME Simulator may be used for both transponder and DME equipment. (*Courtesy IFR, Inc.*)

(Chapter 9) in a single unit. It includes a built-in crystal-controlled signal generator, attenuator, and modulators. The ATC-1200Y3 simulates a ground station in which operating parameters may be varied. It operates in conjunction with an oscilloscope to provide a complete check of essential DME or transponder characteristics. A circulator is used to direct transmitter power to a dummy load and power monitor, and to direct the signal generator to the DME or transponder receiver. Rf monitors are provided for measuring transmitter power and receiver input. A diode switch modulates the signal generator to obtain over 80 dB on/off ratio interrogation.

The ATC-1200Y3 is housed in a cabinet $16\frac{3}{4}$ inches wide by $7\frac{1}{2}$ inches high by $18\frac{3}{8}$ inches deep. Adapters are available for mounting in a standard 19-inch NEMA rack. The unit weighs 36 pounds.

9

Radar Beacon
Transponders

The radar transponder is an outgrowth of systems used during
World War II for making positive identification of friendly aircraft.
The system became known as *Identification—Friend or Foe*, or *IFF*.
Basically, the IFF system consists of a beacon (transponder) in the
airplane. The beacon responds to interrogations on one frequency by
transmitting a reply on another in the form of pulse groups, which
can be coded in many ways. During the war, a designated pulse
group was used to indicate a friendly aircraft. Great efforts were
made for several years to keep IFF secret, although its simple principles were generally known. The IFF system used in World War II
was never completely reliable in military campaigns, but with the
development of the Mark X IFF system after World War II, airborne beacon transponders showed promise of the reliability and
economy necessary for civil use.

The purpose of the transponder is to facilitate air traffic control by
radar. Primarily, it reinforces the signal on the radar scope and provides positive identification of a radar pip. This has the effect of extending radar coverage, and enables air traffic controllers to more
effectively hand over traffic, vector to avoid collision courses, locate
lost aircraft, and maintain close altitude separation. A secondary
function of the transponder is to automatically transmit altitude and
other flight information.

PRINCIPLES OF SURVEILLANCE RADAR

In commercial aviation, ground-based surveillance radar is proving to be the greatest single aid to air traffic control. Before discussing the transponder itself, it is necessary to summarize the radar system with which it operates.

A block diagram of a basic radar system is shown in Fig. 9-1. A magnetron develops a large pulse of rf energy (in some cases, over a million watts peak) which is beamed from the antenna. The receiving circuits are then switched over to the same antenna and receive all resultant echos. The magnetron and receiver circuits are electronically switched between the antenna by "TR" tubes in the preselector. When ionized by the magnetron pulse, these gaseous tubes effectively disconnect the receiver from the rf path in order to prevent pulse energy from reaching the sensitive receiver circuits. The tubes deionize when the magnetron pulse ends.

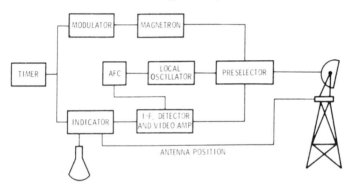

Fig. 9-1. Simplified block diagram of a radar system.

Radars measure the distance to an object by noting the time lapse between the pulse and its echo. The timer is therefore an important part of the system. It usually comprises a crystal oscillator combined with countdown multivibrators or blocking oscillators. The timer circuit controls the rate at which pulses are transmitted, and also motivates the radar indicator, or scope.

The plan position indicator (PPI) presents the operator with a maplike view of the space surrounding the radar antenna. To understand the action of the radar indicator, consider the simplified block diagram in Fig. 9-2. A crystal oscillator operates at a frequency of 8088 Hz. The time required for one cycle at this frequency is 123.7 microseconds. This is also the time it takes for a pulse to travel 10 miles to a target and for its echo to return. Each cycle of the oscillator is shaped into a pulse and fed to a countdown multivibrator.

From here the signal is fed to a sweep generator, which develops a sweep exactly 40 miles long. The trigger countdown provides one pulse for every 20 from the crystal timer. The trigger countdown thus generates about 440 pulses per second, which is the pulse repetition rate of the radar. The cathode-ray tube circuitry is so arranged that the sweep begins in the center of the tube and moves outward. Pulses formed from the 8088-Hz oscillator are made to intensify the beam, which is normally blanked. Thus, four dots will appear on the scope—one in the center (the main pulse), and one at each of the three 10-mile points out to the edge of the scope.

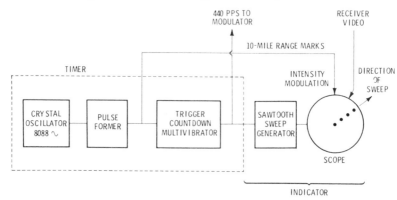

Fig. 9-2. Simplified block diagram of radar timer and indicator.

The radar beam is made to deflect in the direction in which the antenna is pointed. Therefore, with the antenna steadily rotating at 5 to 40 revolutions per minute, the four dots on the radar scope (Fig. 9-2) will rotate at the same rate. Because of its large amount of persistence, the cathode-ray tube screen will continue to glow after the electron beam passes a point on the scope. The moving dots will therefore draw concentric circles that indicate range, and the observer will now "see" a plan view of the space around the radar antenna (Fig. 9-3). A target appears as a bright "pip" that remains visible because of scope persistence, and is renewed each time the sweep passes. Lines can be electronically projected on the scope face to mark important points for the traffic controller. A switch can be provided that allows the scale of the PPI to be expanded if required. The antenna transmits and receives along a very narrow beam, and its angle with the ground can be varied if necessary. At lower elevations, returns from terrain and objects at the shorter ranges produce a scope indication called *ground clutter*. Moving target indication (MTI), on modern radars, eliminates ground clutter by allowing only moving objects to appear.

204

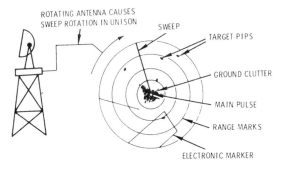

Fig. 9-3. Plan position indicator display.

PRINCIPLES OF TRANSPONDERS

Energy from the radar pulse falls off as the square of the distance traveled. The resulting weaker returns are accompanied by noise, which appears on the screen as countless momentary dots of light called *snow*. Increasing the rf gain of the receiver to compensate for weak returns only results in more noise and may even obscure targets. For positive traffic control it is necessary that targets at maximum ranges be "painted" brightly on the radar screen. The following will illustrate the great advantage of the transponder in reinforcing the radar return.

With ordinary surveillance radar, energy traveling to the target varies inversely as the square of the distance; hence:

$$\text{Energy reaching target} = (K) \frac{\text{energy transmitted}}{(\text{distance})^2}$$

where K is a factor based on antenna characteristics, atmospheric attenuation, etc. Now, the echo that returns to the radar again varies inversely as the square of the distance; therefore:

$$\text{Energy returned} = K^1 \frac{(K) \dfrac{\text{energy transmitted}}{(\text{distance})^2}}{(\text{distance})^2}$$

where K^1 is a factor based on the size of radar target. Or, more simply:

$$\text{Energy returned} = (KK^1) \frac{\text{energy transmitted}}{(\text{distance})^4}$$

From this, we see that the strength of the echo received from a target varies inversely as the *fourth power* of the distance to the target. With the transponder, however, a new pulse is generated upon receipt of a radar interrogation; and this pulse makes a one-way trip

205

from the aircraft to the ground. Thus, the energy of the transponder reply varies inversely only as the *square* of the distance between the airplane and the ground station. This represents a gain of 20 dB, providing a good return at all ranges under all conditions.

The ground-station equipment used with a transponder is known as an *interrogator* and is separate from the radar. The interrogator antenna is mounted on the radar antenna and turns with it. The standard frequency of interrogation is 1030 MHz, and the transponder in the airplane replies at 1090 MHz. The beam width of the interrogator antenna is greater than that of the associated radar, and the transponder uses an omnidirectional antenna. When the radar

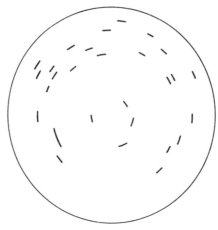

Fig. 9-4. Transponder replies on the radar scope of the ground station.

"pulses," the interrogator simultaneously transmits a pulse group. The transponder returns a group of coded pulses, which are decoded at the ground station. A single pulse is then sent through the radar video and appears on the PPI scope as a short line. The scope trace is deliberately delayed to compensate for the time delay that resulted from the decoding process. Therefore, the transponder trace appears on the PPI scope at the correct location. Because of the relatively greater beam width of the interrogator, the transponder will reply slightly before and after the radar antenna is pointed directly at the airplane, resulting in the short line seen on the scope. Aircraft transponder traces appear on the scope as in Fig. 9-4.

For the moment, the interrogator can be considered as sending a 2-pulse group to challenge the transponder in modes A, B, C, and D, as shown in Fig. 9-5. The spacing between pulses determines the information that the transponder will return. When a transponder is interrogated with the two pulses in modes A or B, it will reply with two to eight accurately spaced pulses 0.4 microsecond in duration and 2.9 microseconds apart. When a transponder is interrogated in modes C or D, the reply pulse will consist of 21- and 25-microsecond frames comprising five to fifteen 0.4-microsecond pulses spaced 1.45 microseconds apart. Mode A is the only mode used at present in the U.S. for identification purposes. Both modes A and B will probably be used eventually to convey type of aircraft, flight number, and other such information to air traffic control.

Mode C is used for automatic altitude reporting. The system utilizes an aneroid, or pressure-type, altimeter coupled to the transponder that automatically transmits altitude in 100-foot increments when the transponder is interrogated in mode C.

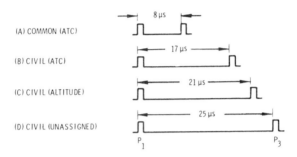

Fig. 9-5. Interrogator modes.

FALSE TARGETS

Ground radars of ATC constantly sweep the sky with their 1030-MHz pulses. All transponders within range of the signal are interrogated. As air traffic increases in a given zone, the interrogations and replies present in that zone become enormous. An en-route ATC radar interrogator will elicit at least 30 replies from a transponder in a given sweep; a terminal (airport) radar interrogator will generate at least 12 replies per pass. Each reply occupies 20.3 seconds, and the first pulse of a train will be 3.3 nautical miles away by the time the last pulse leaves the replying aircraft. Thus, if two aircraft happen to be within 3.3 nautical miles of each other during interrogation, their reply chains will overlap and delude the interrogator. False codes so generated are known as *garbling*. Obviously, garbling increases as traffic increases. Unfortunately, this latter condition is the time when transponders are needed the most.

Reflections of the interrogating pulse are another cause of interrogator delusion. The interrogator beam may be reflected from terrain objects or large structures, causing a transponder to falsely reply to reflections, thus creating the appearance of a target on the screen where none exists. False targets from reflection can be eliminated by utilizing the principle that a reflected interrogation will always travel farther than a direct interrogation. Improved side-lobe suppression (SLS) circuitry in interrogators will correct reflection errors.

Another false target presentation occurs from high-flying jet aircraft, known as *second-time-around* targets. The line-of-sight range of high-flying targets may be such that replies may occur from targets beyond the normal 200 nautical mile range of the radar scope. The reply now arrives after the next consecutive sweep has started. Thus, a reply from an aircraft 230 nautical miles away will appear at 30 miles on the radar scope.

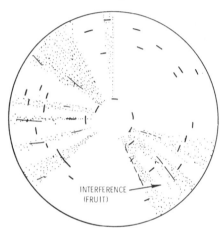

Fig. 9-6. "Fruiting" interference.

The transponder antenna mounted on the airplane is omnidirectional; hence, its energy will reach any station within range. A form of interference known as *fruiting* can result when the transponder reply pulses meant for one station are picked up by another. Since the pulses of different radars are related only at random, the interference will appear as random bright dots (Fig. 9-6). Obviously, the more aircraft there are in an area covered by several interrogators, the worse the interference will be. Fruiting must be eliminated in the video circuitry of the ground interrogating station.

Problems such as fruiting and false targets can be partially eliminated by reducing interrogator power. The important thing here is that receiver sensitivity must be kept at peak rated values or replies will not occur at maximum rated range. Most transponders are equipped with a switch to lower sensitivity when flying close to interrogators. A serious problem is that busy pilots forget to return the transponder to high sensitivity and vice versa.

Experience shows that as the transponder population increases, a point of saturation is reached. It is not now possible to predict the true point of saturation; however, it may be expected that this will result in great economic impact. It means that at some unknown future date, the adoption of a radically different system of electronic aircraft identification will be required. A system now under development, the Discrete Address Beacon System (DABS), will enable controllers to communicate with aircraft on the radar screen by data link. A readout device will display data to the pilot with potentially up to 16,777,216 code combinations. Interrogation from remote data links will eliminate fruiting.

SIDE-LOBE SUPPRESSION (SLS)

The interrogator radiates from an antenna that is attached to and rotates with the main radar antenna. The interrogator antenna is principally unidirectional, but will have side lobes as shown in Fig. 9-7. At ordinary distances, the transponder replies only when the main radar antenna is pointed at it. But when the target is in close, the transponder also replies to side lobes of the interrogating antenna, resulting in false indication. Systems of side-lobe suppression have been devised to eliminate such spurious interrogation.

The International Civil Aviation Organization (ICAO) has standardized the so-called 2-pulse side-lobe suppression system (Fig. 9-7A). The first pulse (P_1) of each 2-pulse group from the interrogator is radiated from an omnidirectional antenna, and the second pulse (P_2) follows from the directional antenna. The omnidirectional pulse is designed to appear at least 10 dB more than the radiation from any side lobe. The transponder is designed to reject all pulse

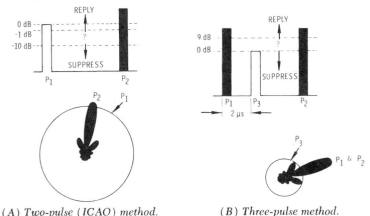

(A) Two-pulse (ICAO) method.   (B) Three-pulse method.

Fig. 9-7. Side-lobe suppression.

pairs in which the first pulse appears larger than the second; therefore, it will not reply to side lobes.

A directional antenna concentrates much more energy in a given space than an omnidirectional antenna. This is one objection in the system just described—the large amount of power required for the omnidirectional radiation. After studying the matter, the FAA adopted a 3-pulse sidelobe suppression system (Fig. 9-7B). Both interrogating pulses of the group (P_1 and P_2) are radiated from the directional antenna; but another pulse (P_3), added between the two, is radiated from an omnidirectional antenna and is amplitude regulated. The system is adjusted so that when omnidirectional pulse P_3 reaches a level above that of the most powerful side lobe that can be expected, the transponder will not reply.

In the 2-pulse method, the omnidirectional power must be almost as high as that radiated from the directional antenna. It can be seen that much less power need be radiated from the omnidirectional antenna in the 3-pulse method. The time interval between the two interrogating pulses of the ICAO system is the same as that between the interrogating pulses of the FAA system, and both use the same intervals for modes. Hence, a transponder designed for the ICAO system will accept interrogations from an FAA ground station. However, since one system will not operate with the side-lobe suppression of the other, there will be spurious responses to side lobes. For international air fleets, most manufacturers produce a modified transponder that incorporates both 2-pulse and 3-pulse side-lobe suppression and is mutually compatible without switching or adjustment.

CODING AND DECODING

Decoding is accomplished by coincidence circuitry. A simple coincidence circuit, using only a diode, is shown in Fig. 9-8. The input enters the delay line, but does not immediately appear at point C because of the diode polarity. After emerging from the delay line, the pulse appears at points B and C. Upon cessation of the input pulse, point C is more positive than point A, so the diode con-

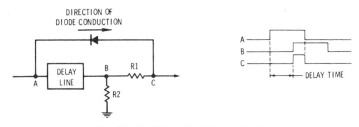

Fig. 9-8. A diode coincidence circuit.

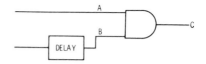

Fig. 9-9. A basic side-lobe suppression circuit.

ducts, dropping voltage across R1 and causing point C to revert to ground potential. Side-lobe suppression may also be achieved as shown in Fig. 9-9.

The original system of transponders envisioned only 64 mode-A identity codes. The standard system now adopted by international agreement, however, includes 4096 mode-A identity codes plus the same number of mode-C codes for automatic altitude reporting. Transponders now being produced have this expanded code capability.

A transponder must be able to return a pulse train in accordance with the code assigned by ATC. The original 64 mode-A identity code system is shown in Table 9-1. This system is still used by air-

Table 9-1. Pulse-Train Code

F_1 A_1 A_2 A_4 B_1 B_2 B_4 F_2 SPI

Code No.	Information Pulses Present	Code No.	Information Pulses Present	Code No.	Information Pulses Present
00	None	26	$A_2B_2B_4$	53	$A_1A_4B_1B_2$
01	B_1	27	$A_2B_1B_2B_4$	54	$A_1A_4B_4$
02	B_2	30	A_1A_2	55	$A_1A_4B_1B_4$
03	B_1B_2	31	$A_1A_2B_1$	56	$A_1A_4B_2B_4$
04	B_4	32	$A_1A_2B_2$	57	$A_1A_4B_1B_2B_4$
05	B_1B_4	33	$A_1A_2B_1B_2$	60	A_2A_4
06	B_2B_4	34	$A_1A_2B_4$	61	$A_2A_4B_1$
07	$B_1B_2B_4$	35	$A_1A_2B_1B_4$	62	$A_2A_4B_2$
10	A_1	36	$A_1A_2B_2B_4$	63	$A_2A_4B_1B_2$
11	A_1B_1	37	$A_1A_2B_1B_2B_4$	64	$A_2A_4B_4$
12	A_1B_2	40	A_4	65	$A_2A_4B_1B_4$
13	$A_1B_1B_2$	41	A_4B_1	66	$A_2A_4B_2B_4$
14	A_1B_4	42	A_4B_2	67	$A_2A_4B_1B_2B_4$
15	$A_1B_1B_4$	43	$A_4B_1B_2$	70	$A_1A_2A_4$
16	$A_1B_2B_4$	44	A_4B_4	71	$A_1A_2A_4B_1$
17	$A_1B_1B_2B_4$	45	$A_4B_1B_4$	72	$A_1A_2A_4B_2$
20	A_2	46	$A_4B_2B_4$	73	$A_1A_2A_4B_1B_2$
21	A_2B_1	47	$A_4B_1B_2B_4$	74	$A_1A_2A_4B_4$
22	A_2B_2	50	A_1A_4	75	$A_1A_2A_4B_1B_4$
23	$A_2B_1B_2$	51	$A_1A_4B_1$	76	$A_1A_2A_4B_2B_4$
24	A_2B_4	52	$A_1A_4B_2$	77	$A_1A_2A_4B_1B_2B_4$
25	$A_2B_1B_4$				

craft equipped with 64-code transponders. However, for ATC to utilize one or a combination of the 4096 discrete codes including the 64 basic select codes, 4-digit code designation is used. For example, code 2100 is expressed as TWO ONE ZERO ZERO. Pilots flying an aircraft equipped with a 64-code transponder disregard the last two numerals of the code issued by ATC. That is, if they are assigned code 2100, they only set in the numerals "21."

The assigned reply code is set on the transponder control unit in the cockpit. In Table 9-1, F_1 and F_2 are framing pulses. A_1 through B_4 are information pulses spaced at 2.9 microseconds. For example, code 62 (or 6200) would consist of framing pulses F_1 and F_2, which are always present, and pulses A_2, A_4, and B_2. SPI is the identification pulse. As another example, code 20 (or 2000) would consist of F_1 and F_2 plus pulse A_2.

A typical transponder control head is shown in Fig. 9-10. The large control knob in the center contains the following positions: OFF, STBY, ON, LO SENS, and TEST. In the OFF position, the transponder is turned completely off. In STBY (standby) position, the transponder is on but is not replying to ground interrogations. In the ON position, the transponder is on and replies in mode A to all ATC interrogations. The LO SENS position reduces the sensitivity of the receiver, allowing replies only to local stations, thus eliminating some fruiting interference and overloading. The four code-selector switches at the top are set to the assigned reply code. When the pilot depresses

Fig. 9-10. A typical transponder control head. (*Courtesy NARCO Avionics, a Division of NARCO Scientific Industries, Inc.*)

the IDENT button on request from ATC, the unit will reply with identification (SPI) pulses for a preset length of time. Code-C automatic altitude reporting is activated by switching the ALT switch to ON. The unit continues to reply to mode-A interrogations while also being interrogated in mode C. The TEST position allows the pilot to check the operation of his receiver whenever desired or deemed necessary. The MON (monitor) light comes on whenever the unit is being self-tested, when the unit replies to ground interrogations, and whenever the IDENT feature is in operation.

ALTITUDE REPORTING

Airplanes that have transponders equipped for altitude reporting are much easier to control on IFR flights, since ATC is able to display the actual altitude directly at the controller's console. Almost all transponders now produced have provisions for altitude reporting. Fig. 9-11 shows a Bendix encoding altimeter for use with transponders. This instrument meets the following specifications:

Conformance Specifications—Altimeter-TSO-C10b; Encoder-TSO-C88.

Range——1000 feet to 20,000 feet.

Altimeter Accuracy—0 feet altitude ±20 feet; 10,000 feet altitude ±80 feet; 20,000 feet altitude ±130 feet.

Altimeter/Encoder Accuracy—Within ±75 feet of required transition points at all altitudes.

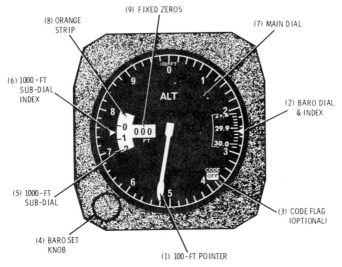

Fig. 9-11. Bendix encoding altimeter for reading up to 20,000 feet.

Power Requirement—Encoder and optional warning flag; 14 or 28 volts dc, 4.5 watts (max).

Size—3.12 inches in diameter plus baroset sump (3 inches round) by 5.5 inches long behind panel.

Weight—2.20 pounds.

The Bendix altimeter operates in the same manner as the common instrument ordinarily used, but is also capable of supplying digital signals to a transponder (Fig. 9-11). The pointer (1) is read directly against the main dial (7). One full revolution is equal to 1000 feet. The 1000 foot sub-dial (5) is read against the index marker (6) and displays from −1000 to 20,000 feet. At least three numbers are always visible when ascending or descending and dials are thus in motion. Below zero sea-level altitude an orange strip (8) is visible. Three fixed zeros (9) are provided at all times. The 1000 foot sub-dial moves in direct proportion to the pointer. The rule for reading this instrument is: *Read the smallest dial number next to the index, plus the pointer.*

An altimeter must have a means of correcting for differences in atmospheric pressure due to other than altitude change. It will be remembered that areas of low and high pressure are always present as weather patterns change. Standard sea-level air pressure is assumed to be 29.92 inches of mercury (in Hg).° When pressure at a given locality differs from this, the correct value must be set into the altimeter. Local altimeter settings are radioed to the pilot as the flight progresses. These are set into the baro dial (2), causing the instrument to read actual altitude above sea level.

The altitude that will be sent by the transponder, however, will not be corrected for barometric pressure. One reason for this is that the true compensation for barometric pressure at the altitude and location of an airplane can never be known with certainty. Barometric compensation is known most accurately during the landing, fortunately, since the field elevation and barometric compensation at the airport can always be known exactly. However, if each of two airplanes has its altimeter set to 29.92 in Hg, regardless of actual atmospheric pressure, and if the two are at the same location and at the same altitude, the readings on the respective altimeters will be the same even if the indicated altitude is incorrect. Thus, for collision avoidance, airplanes should report altitude without compensation for barometric changes.

°Under the new International System of Units, commonly known as the SI metric system, inches of mercury (in Hg) as a unit of atmospheric pressure is replaced by the pascal (Pa); i.e., 1.0 in Hg (32°F) = 3.386 kPa; or, 29.92 in Hg (32°F) = 101.325 kPa.

The Bendix unit features an optical encoder that is driven by an aneroid mechanism. The optical pickoff eliminates friction between the aneroid mechanism and digitizing device. The digital output of altimeters for use with transponders is standardized as 10-bit parallel data. Eleven wires are used. Standard lead designations between encoding altimeter and transponder are shown in Fig. 9-12.

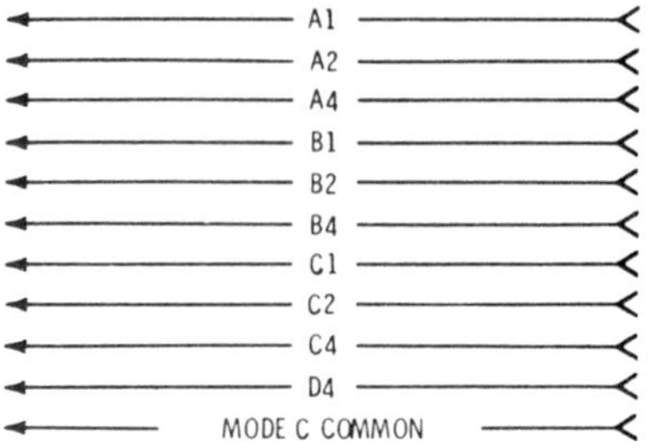

Fig. 9-12. Standard lead designations between encoding altimeter and transponder.

A1	
A2	
A4	
B1	
B2	
B4	
C1	
C2	
C4	
D4	
MODE C COMMON	

THE BASIC TRANSPONDER SYSTEM

The rf and if portions of an ATC transponder closely resemble the DME interrogator, as can be seen from the block diagram of Fig. 9-13. The ground station always transmits at 1030 MHz and the transponders reply on 1090 MHz. Thus, with an if of 60 MHz, a single stabilized oscillator serves for both the local oscillator and transmitter. The preselector prevents interaction between the transmitter and receiver. The detected and amplified video pulses are fed to a spike eliminator, which blocks all impulses of less than 0.3 microsecond in order to prevent action from random noise pulses. The decoder is set to respond to the selected interrogation mode (that is, spacing between the first and last interrogating pulses). At the same time the pulse enters the decoder, it also enters the side-lobe suppression circuit, which is actually an amplitude-selective decoder responding only to pulses from side lobes. The output of the decoder is fed to the keyer, where a pulse is produced. This pulse, in turn, causes the encoder to form a series of pulses according to the selected code.

The automatic overload control counts the number of keyer pulses per second, and reduces the sensitivity of the if strip if more than 1600 pulse groups per second are received. Also, with certain reply codes, it is possible for the transponder to produce many more pulses than it receives. Eventually the average power will increase to the point where it overloads the transmitter. To prevent this, the automatic overload control also counts the pulses from the encoder; and when the average power exceeds a certain amount, the if sensitivity is again reduced. Despite the lower sensitivity, side-lobe sup-

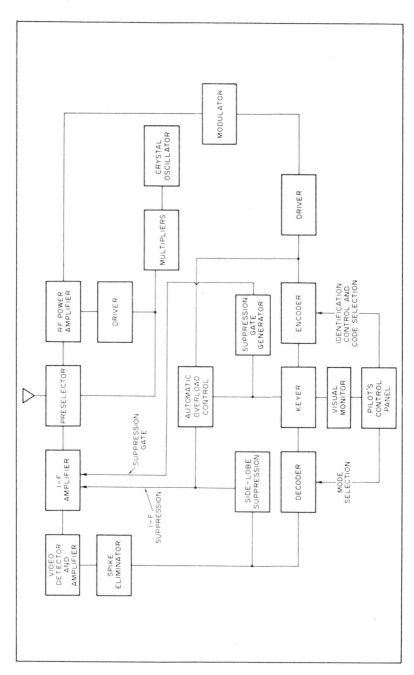

Fig. 9-13. Simplified block diagram of a transponder.

216

pression will still allow the transponder to reply to the stronger main pulses of closer interrogators—the more important from the standpoint of air traffic control.

In accordance with the pilot's code selection, the keyer causes the encoder to set up a train of pulses that trigger the modulator, which forms the rf pulses that ensue from the rf power-amplifier stage. The keyer also causes the generation of a suppression gate pulse, which disables the receiver while a pulse train is being transmitted.

Encoding the return signals is now accomplished with ICs, resulting in lighter weight and trouble-free performance. An example is the encoding circuitry of the Genave BETA/5000 Transponder shown in Fig. 9-14. The control panel of this unit is shown in Fig. 9-15. This modern lightweight unit meets TSO for all operational altitudes and may be coupled to any ARINC standard encoding altimeter. The identification code requested by ATC is set on the digital dial.

Pulses are originated at the clock circuit. The clock is a ringing oscillator formed by L102, C178, and C179. The oscillator is tuned to ring at the correct frequency to time the reply sequence. The encoder enable gate initiates oscillation. A feedback path through R189 enhances oscillation. There are two outputs of the clock; each is 180° out of phase with the other. These are fed to the sequential pulse generator.

Inputs to the sequential pulse generator are applied from the encoder enable line and each phase of the clock output. The encoder enable gate initiates sequencing. The clock outputs time the sequence.

The basic sequential pulse generator circuit consists of two inverters with appropriate input circuitry and dc coupled outputs (Fig. 9-16).

When a negative-going input is applied to C1, the input to IC1 goes negative, then returns to quiescent level in the time determined by the value of C1 and the input impedance of IC1. When the input to IC1 goes negative, the output is prevented from going positive by the clamping action of IC2. The output of IC1 will go positive unless clamped down by the output of IC2. If the input to IC2 (first clock pulse) goes negative at the same time as the input to IC1, the outputs of both ICs will go positive and remain positive until one of the inputs returns to its quiescent state. This will generate a positive pulse at the IC outputs that will be used to trigger the reply pulse. When this output pulse falls, a negative input will be applied to IC3, the first IC in the second circuit, through C2. The second circuit functions in the same manner as the first. The input of IC3 goes negative from the trailing edge of the output pulse of the first circuit. As the input to IC3 goes negative, the second phase

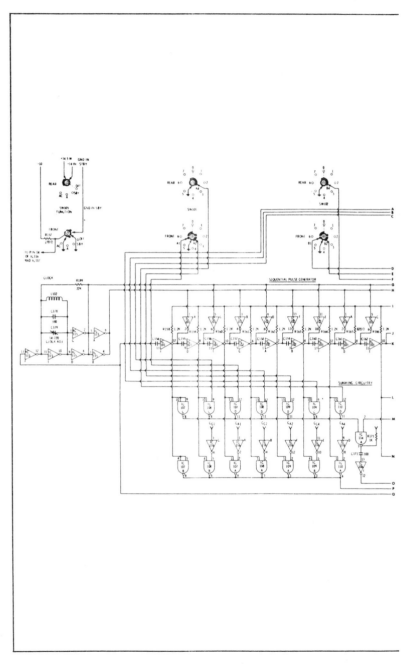

Fig. 9-14. Encoding circuitry of the Genave BETA/5000 transponder.

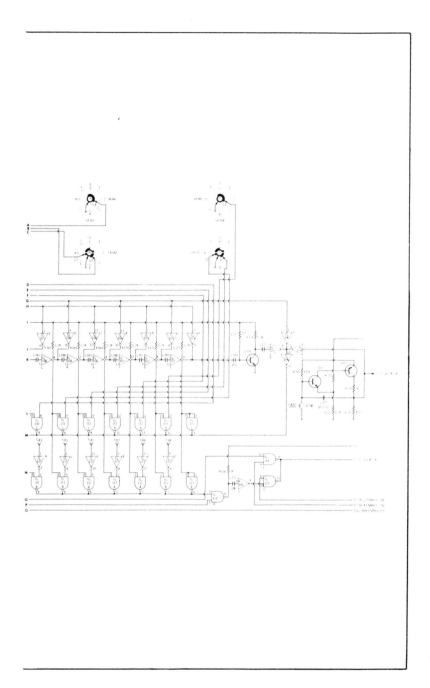

(Courtesy General Aviation Electronics, Inc.)

clock output will also go negative allowing the outputs of the second pair of ICs to go positive.

Fifteen circuits such as this are used in the sequential pulse generator. A similar circuit with an additional IC, requiring an additional negative output is used in the IDENT pulse sequencing circuit.

Complete output timing for the sequential pulse generator is shown in Fig. 9-17.

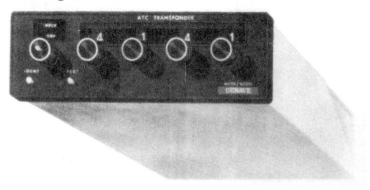

Fig. 9-15. The Genave BETA/5000 transponder. (*Courtesy General Aviation Electronics, Inc.*)

The IDENT timer (right side of Fig. 9-14), consisting of Q102, Q103, and associated circuitry, operates as a triggered one-shot multivibrator that provides a negative pulse of approximately 20 seconds duration when the IDENT switch, SW106, is depressed. The switch causes C177 to discharge, turning off Q102 and Q103. When the switch is released, C177 recharges through R209, turning Q102 and Q103 on. This takes approximately 20 seconds, after which the input of IC106C goes positive.

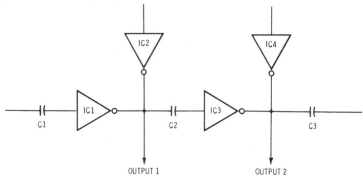

Fig. 9-16. Basic circuit of sequential pulse generator in Genave BETA/5000 transponder.

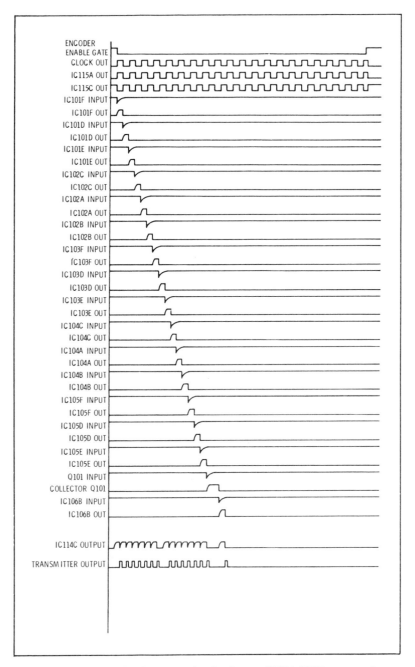

Fig. 9-17. Sequential pulse timing for the Genave BETA/5000 transponder.

The pulse sequence output is applied to the summing gates of IC107 through IC113. The Mode A and Mode C summing gates operate in the same manner. One input to each IC gate is fed from the sequential pulse generator. The other is connected to either the mode selector switches (Mode A) or the altitude digitizer (Mode C). The switched input to each gate is switched to ground unless an output for that respective pulse is desired.

Outputs from both the Mode A and the Mode C summing gates are fed to gates IC114A and IC114B, respectively. The summed outputs of these two gates are each integrated (C171 and C172) and fed to an inverter. When the output of a gate goes negative the output of the inverter will go positive.

The positive pulses from the inverters are applied to two reply-enable gates, which determine whether a Mode A or Mode C reply will be transmitted. IC 114C is the Mode A reply-enable gate. When a Mode A Enable is sent from the decoder, pin 9 of IC114C goes positive for 60 microseconds permitting the Mode A reply train to pass through the gate. When a Mode C Enable is sent to IC114D, the Mode C reply-enable gate, pin 12 will be positive for the duration of the Mode C Enable, allowing the Mode C reply to pass. From the reply-enable gates, the assembled pulse reply trains are sent to the pulse-shaper circuitry, then applied to the modulator.

TRANSPONDER MAINTENANCE AND TESTING

Transponders have proven quite reliable in service. The military APX-6 transponders used with the Mark X IFF system had a good record of reliability, once properly installed and checked. Troubleshooting the newer transponders can often be accomplished by merely substituting modules until the defective section is located by the process of elimination. An inoperative transponder will be revealed by the absence of a visual monitor or a complaint from air traffic control. It is good practice, before removing the transponder, to check its shock mounting and its interconnecting cables.

At overhaul centers, it is necessary to test all characteristics to determine whether specifications have been satisfied. This includes a check of the following:

Receiver bandwidth and sensitivity.
Image response.
Receiver decoder tolerance.
Transmitter trigger (PAR).
Reply frequency.
Transmitter pulse characteristics.
Transmitter power output.

Fig. 9-18. The Collins 578X-1 ATC transponder bench test set.
(*Courtesy Collins Div., Rockwell International*)

Identification time.
Transponder delay time.
Reply pulse position and jitter.
Receiver output.
Suppression input and output.
Automatic output control (AOC).
Discrimination of interrogation-pulse amplitude differences.
Echo suppression and recovery times.
Random triggering rate.
Transponder "dead" time.

These characteristics can be tested by using the Collins 578X-1 ATC transponder bench test set (Fig. 9-18) in conjunction with the Boonton 235X Navigation Aid test set in Fig. 9-19, and a suitable oscilloscope.

The IFR, Inc., Model ATC-600A transponder/DME ramp test set (Fig. 9-20) is designed for ramp or bench testing. Using radiated signals, this unit will perform all tests required by FAA regulations 91.177 and 43, Appendix F, as amended December 19, 1973. The unit weighs 19 pounds and operates up to two hours on internal nickel-cadmium batteries. It measures 11 inches wide by 4 inches high by 12 inches deep. A built-in charger operates whenever the unit is connected to an ac line. When used on the ramp, a tripod-mounted antenna system is coupled to the aircraft antenna, as shown in the sketch in Fig. 9-21.

Fig. 9-19. The Boonton 235X Navigation Aid test set.
(*Courtesy Boonton Electronics Corp.*)

Transponder tests allow quick determination of transmitter power, frequency, percent reply, pilot's code, and encoded altitude. Variable-spaced interrogations and reply framing timing checks are

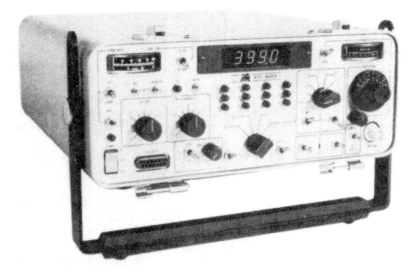

Fig. 9-20. The IFR, Inc., Model ATC-600A transponder/DME ramp test set.
(*Courtesy IFR, Inc.*)

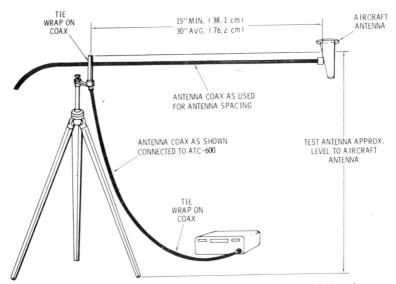

Fig. 9-21. The ATC-600A transponder/DME test set is coupled to the aircraft antenna on the ramp as shown here.

also provided. DME checks can be made on the aircraft at the same time that ramp tests are run on the transponder. DME testing capabilities include accurate range and velocity simulation, power and frequency, and PRF measurements. Characteristics of the ATC-600A are as follows:

TRANSPONDER

Interrogations Output
Mode: A/C, Altitude or Pilot Code, 2:1 interlace, or mode A (*B mode available on request*).

Pulse Spacing: P_1–P_3 variable ±1 μs from nominal for input decoder gate tests.

PRF: 235 Hz.

SLS Test: P_2 inserted at 0 dB or −9 dB relative to P_1.

Power: −66 to −79 dBm direct with 34 dB pad or radiated with properly spaced antenna for receiver sensitivity tests.

Reply Measurements
Power: 10 W to 1.5 kW peak, ±3 dB; direct with 34 dB pad or radiated with properly spaced antenna.

225

Frequency:	1086 to 1093 MHz, 0.3 MHz resolution.
Pilot or Altitude Code:	Binary and numerical readout, −1.0 to 126.7 thousand feet.
Percent Reply:	0–100%, either A/C or A mode.
F$_2$ Pulse Position:	Measurement of rising and falling edge ±0.5 μs from nominal.
Status Lamps:	Ident Pulses, Invalid Altitude Code, and No Altitude Code.
Encoder Test:	Direct connection accepts altitude encoder.

DME

Interrogations Measurements

PRF:	0 to 30 and 0 to 300 Hz.
Power:	10 W to 1.5 kW peak, ±3 dB; direct with 34 dB pad or radiated with properly spaced antenna.
Frequency:	1037 to 1044 MHz, 0.3 MHz resolution.

Reply Output

Frequency:	Paired with VOR: 108.000 MHz (X-channel) or 108.050 MHz (Y-channel).
Power:	−45 ±3 dBm; direct with 34 dB pad or radiated with properly spaced antenna.
Range:	0 to 399 miles in 1 mile steps. Accuracy ±0.07 mile.
Velocity:	Crystal-controlled digital velocity with rates of 50, 75, 100, 150, 200, 300, 400, 600, 800, 1200, 1600, and 2400 knots. Inbound or outbound starting from any selected range. Range steps in velocity mode are 0.025 mile (system), 0.1 mile displayed.
Percent Reply:	100% or 50%.
Ident Tone:	1350 Hz with equalizing pulses.

10

Weather Radar

Radar was developed before World War II by several nations working separately, and was kept a highly guarded secret. At the beginning of the war, the United States and Great Britain pooled their knowledge. The result was one of the greatest developments to come out of the war. The word "radar" stands for *RA*dio *Detec*tion *A*nd *R*anging. The first airborne radar was developed for fighter-interceptor and antisubmarine operations. At the close of the war, radar was adapted for civilian use.

In commercial aviation, airborne radar is primarily intended for detecting weather conditions that could lead to undesirable turbulence or hail. Contrary to popular belief of the lay public, commercial airborne radar has no relationship to air traffic control or collision avoidance. As will be pointed out, however, some systems can be used to sense terrain.

Certain classes of commercial carriers are required by FAA regulations to have weather radar installed. Weather radar operates on the principle that most forms of turbulence are accompanied by precipitation which reflects radio energy. By observing the patterns of precipitation on a radar screen, the pilot can avoid thunderstorms and other squall areas, and can thus cross weather fronts at points of least turbulence. Radar can also display a "map" of the terrain beneath the airplane which can be used for navigation.

PRINCIPLES OF WEATHER RADAR

Weather radar operates on the same principle as the radar system discussed in the previous chapter. Recall that very short pulses are radiated from the antenna in a narrow beam. If the pulses reach a large object such as an airplane, comparatively large echo pulses will be returned. Should the pulses travel through precipitation, a certain amount of energy will likewise be returned from each droplet. When the operating frequency of the radar transmitter is increased, even more energy will be returned by the precipitation. As each pulse is generated, a beam of electrons is swept outward from the center of the radar scope, in the direction the antenna is pointing at that instant. Since the radar antenna is rotating, the beam on the scope is rotating in accordance with it. Through appropriate receiver and video circuitry, the returning energy intensifies the beam of the scope. Since many pulses per second are transmitted from the rotating antenna, a pattern of precipitation is thus "painted" on the face of the scope as the electron beam passes. The face of the cathode-ray tube remains illuminated for a brief instant afterward, and thereby provides an overall picture.

Thunderstorms are traditional hazards to aerial operations. They are formed when warm moist air collides with dry cold air at high altitudes. They develop vertically in large cumulus clouds, usually between 8000 and 15,000 feet, but have been known to extend up to the stratosphere. In general, the higher the storm reaches, the more violent it is. At 30,000 feet, hail can generally be expected. A storm reaching to 50,000 feet indicates the possibility of tornadoes.

Rainfall gradient is defined as the variation of the rate of rainfall with distance. Violent turbulence is present in areas where rainfall gradient is greatest. It is possible to outline areas of greatest gradient by means of special video circuitry known as *contour display*. Steep gradients are displayed on the indicator scope as dark spots, or cells, surrounded by a ring of bright returns (Fig. 10-1). When contour circuitry is switched on, the absence of dark cells usually means no severe turbulence is present. Fig. 10-2 shows a sequence of radar displays as an aircraft proceeds to avoid two areas of heavy turbulence (the dark cells). Heavy rainfall not accompanied by turbulence is revealed by the solid bright areas. Light, scattered, fuzzy returns represent gradual changes in the rainfall and are not detrimental to flight. Hail is formed by updrafts or downdrafts within turbulent cells, after which the hail may be ejected from the cell itself. Hail usually appears on the scope as fingers, scallops, or U-shaped projections extending from areas of intense echoes.

Since weather is a dynamic thing, storm cells and precipitation patterns are constantly changing. The average life of a storm cell is about 1½ hours. Thus, a pilot can make decisions which, based on aircraft speed and other factors, can determine the procedure for entering storm areas. Together with a knowledge of meteorology, the pilot can correctly interpret radar weather displays and thus be warned of rain, snow, turbulence, and even conditions that might lead to icing or poor visibility. For example, the upper areas of a storm can be explored by tilting the antenna. If the temperature above a rain area is below freezing and bright returns are received on the radar scope, wet snow is probably present. Often pilots of radar-equipped aircraft are requested to make observations from which weather reports are compiled.

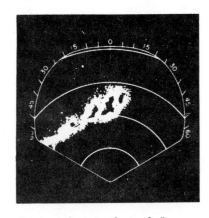

Fig. 10-1. The radar indicator display with "contour" circuitry on. An area of heavy turbulence can be seen dead ahead. (*Courtesy Collins Div., Rockwell International*)

The choice of frequency depends on what is desired from a weather radar system. Higher frequencies provide brighter reflection, but have a lower range because their energy is dissipated more quickly. The two frequency ranges adopted for commercial use are the 5200- to 5900-MHz C band and the 8500- to 10,000-MHz X band.

The peak power of each transmitter pulse determines the ability of a radar system to detect objects. At the same time, the physical size and power consumption of radar equipment depend on its average power dissipation. Average power depends, first, on the peak power and duration of each pulse and, second, on the number of pulses per unit of time. The average power of a radar is therefore the product of the peak power, the pulse repetition rate, and the pulse duration. Reducing the pulse repetition rate below a certain value may result in insufficient average video power being returned. Consequently, targets may be lost as the beam passes them by.

The ability of a radar to distinguish between two separate objects is known as *resolution* and depends on the pulse width. A 1-microsecond pulse represents a train of waves sweeping an area 984 feet wide. With this pulse width, it is obvious that any two objects must be more than 984 feet apart in order to appear as two separate echoes. Thus, the shorter the pulse duration, the better the resolution and the more capable the radar is of displaying terrain features with detail. As a rule, the narrower the pulse, the wider the frequency spectrum it occupies. A 1-microsecond pulse, for example,

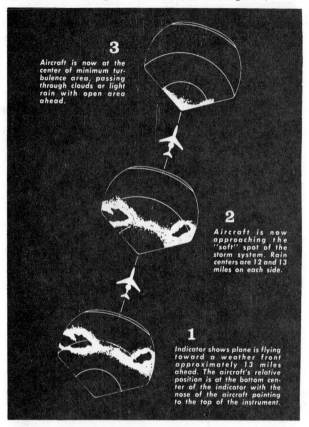

3
Aircraft is now at the center of minimum turbulence area, passing through clouds or light rain with open area ahead.

2
Aircraft is now approaching the "soft" spot of the storm system. Rain centers are 12 and 13 miles on each side.

1
Indicator shows plane is flying toward a weather front approximately 13 miles ahead. The aircraft's relative position is at the bottom center of the indicator with the nose of the aircraft pointing to the top of the instrument.

Fig. 10-2. Radar indications as airplane flies between two areas of heavy turbulence. (*Courtesy Collins Div., Rockwell International*)

occupies a spectrum width of 2 MHz. Hence, the wider receiver if bandpass usually required by the narrower pulse increases the noise level and reduces the range.

As pointed out in Chapter 9, the returned energy varies inversely as the fourth power of the distance. Hence, returns from nearby objects will appear much brighter than those from a distance. Special circuitry known as STC (sensitivity time control) reduces the receiver gain at the start of the trace, and then increases it as the more distant returns are received. Thus, with STC a uniform bright-ness is "painted" on the radar scope for both near and distant objects, and a more accurate display of the terrain is provided.

GROUND MAPPING AND AVOIDANCE
' OF TERRAIN COLLISION

Although not designed specifically for ground mapping, airborne weather radar will clearly indicate principal terrain features, such as cities, lakes, rivers, and shorelines.

The terrain is viewed by depressing the antenna with the TILT control. The PPI display may include an entire circle around the airplane. Since weather radar is mounted in the nose, however, the fuselage will obstruct the view, causing a dead area in the rear. Some radars sweep back and forth only in front of the airplane, utilizing the off-center sweep shown in Fig. 10-1. In either in-stance, the scale of the display can be changed. Weather radar usually has three scales with range marks. A maximum display of 150 miles is sufficient range for flight operations. Energy from the simple parabolic reflector antenna of the weather radar is con-centrated into a narrow beam. When the beam is tilted, only seg-ments of the terrain can be viewed. Fig. 10-3 illustrates the appear-ance of terrain on a weather-radar scope as the airplane flies along a river at approximately 10,000 feet and the antenna is tilted 6° below the horizontal plane. The segment of terrain being scanned is approximately 19 miles wide.

Most displays are quite simple to interpret; others will require considerable skill and experience. Boundaries between land and water offer the most striking contrast, and shoreline flying is often resorted to in parts of the world lacking modern navigational aids. Terrain scatters radar energy in all directions, with part of it being returned to the radar antenna.

Since different terrains and ground objects reflect with different intensities, the gain of the receiver must be variable to give the best display. Even with STC, cities and other built-up areas, particularly groups of tall buildings nearest the airplane, will often appear brighter than more distant objects. Hence, the brighter environs of an approaching city may create the illusion of moving the city closer to the airplane. In general, the image detail increases as dis-tance is decreased. Note that no returns from the periphery of the radar scope does not necessarily indicate the absence of objects, but could mean that the system is incapable of reaching that far.

Weather radar can also help prevent terrain collision in moun-tain areas. The antenna platform, which is kept level by a vertical gyro, can be tilted to reveal terrain at a certain height and distance from the airplane. For example, the indicator in Fig. 10-4 will reveal the segment shown when set to 30 miles. Since the high ground blocks the signals, there are no returns from the area behind it. In the illustration, the antenna is tilted so that when the terrain

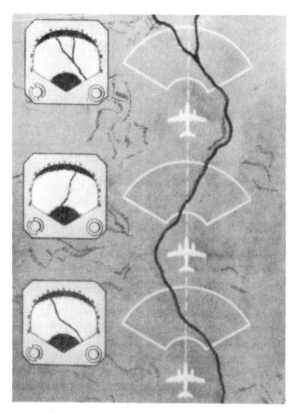

Fig. 10-3. Ground-map interpretation from a radar scope. (*Courtesy Collins Div., Rockwell International*)

appears even with the 10-mile marker, the aircraft will be able to just clear the terrain at the chosen safe height. Appearance of terrain within the 10-mile marker would indicate that the course must be changed immediately.

OTHER METHODS OF STORM AVOIDANCE

Severe turbulence that leads to trouble in flight is almost always associated with thunderstorms, and thunderstorms are accompanied by lightning. If some means were available to continuously detect the direction and distance of lightning, this would be another method of avoiding severe turbulence.

One method of presenting the pilot with storm information from the presence of lightning has been developed.[°] This system is passive; that is, no energy originates in the aircraft system as in radar. Lightning discharges are picked up by a 360° direction finder scanning system. The direction finder is sensitive to energy in the spectrum below 200 kHz. The energy of electrical storms is high in this lower range, and the lower-frequency radiation fields tend to dissipate as a regular function of distance over a long range.

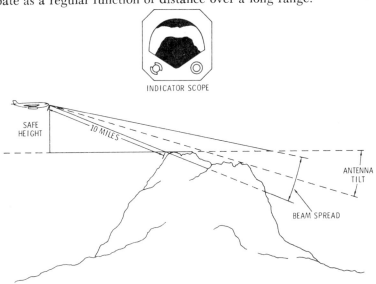

Fig. 10-4. Technique for avoiding terrain collision.

Distance to the lightning discharge is approximately determined by processing the signal strength. The received signals are processed in a "hybrid analog-digital" system. The analog system enhances the sensitivity-distance relationship, and the resulting signal is sent to a digital system that forms a video signal and holds it temporarily.

A lightning discharge is displayed as a bright green dot on the face of a high intensity cathode-ray tube operating in the PPI mode. An area of storm activity is displayed as an animated array of green dots. The intensity of the storm is revealed by the activity and area of the dot clusters. The display may be changed in scale in steps of 20, 50, and 100 nautical miles.

This method offers a lower-cost system that will be of advantage

---

[°]Ryan Model WX-7 Stormscope, Stormscope Co., Columbus, OH.

to lower-cost airplanes. The disadvantage is that it does not reveal precipitation and cannot, therefore, be used to reveal the presence of other weather phenomena. On the other hand, the method provides the pilot with a 360° view of storm conditions, whereas radar can only operate forward from the nose of the aircraft where the radome is located.

UHF TECHNIQUES IN RADAR

The pulses of rf energy for the radar system are generated by a magnetron oscillator. The magnetron, developed at the beginning of World War II, was the first device that could furnish high power at microwave radar frequencies. It is a vacuum tube having a cylindrical cathode surrounded by a slotted anode. The operation of this tube is based on the phenomenon that a moving electron will

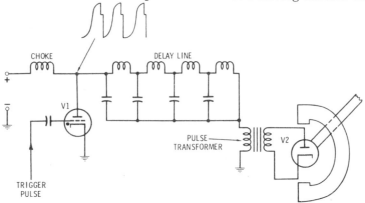

Fig. 10-5. Simplified schematic of a radar modulator.

set up a magnetic field at right angles to its motion. A large permanent magnet surrounding the magnetron furnishes a magnetic field that penetrates the anode. An electron leaving the cathode has its magnetic field acted on immediately by the field of the permanent magnet. As it accelerates toward the positive anode, the electron is acted upon more and more by the magnetic field, and is made to take a helical path to the anode. In so doing, the swarms of electrons set up a wave motion of a frequency determined by the dimensions of the slotted anode and associated waveguide. Power is drawn from the magnetron by either a coupling loop or waveguide.

The magnetron is actuated by a strong, essentially square wave of the proper duration, which has been processed by the delay line in the modulator (Fig. 10-5). In this circuit, thyratron V1 is fired by a trigger pulse and discharges the delay line through the pulse transformer. The delay line has the same characteristic impedance as the primary of the loaded pulse transformer, and the resultant pulse output has twice the duration of the delay time of the line. The choke and the first input capacitor of the delay line are chosen to resonate at a frequency near the pulse repetition rate of the system. The waveform across thyratron V1 is as shown.

The local oscillator of a radar system must be a device capable of emitting uhf in the range of the magnetron. It must also be capable of automatic tuning to follow magnetron drift. The reflex klystron—an electron tube that consists of a cathode, focusing electrode, cavity resonator, and repeller electrode—has been in use for

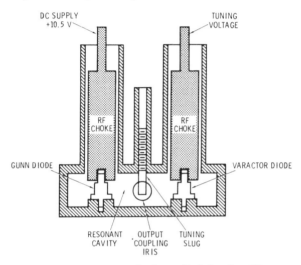

Fig. 10-6. Cross section of a Gunn diode local oscillator.

a long time. Electrons drawn from the cathode are formed into a beam and accelerated toward the anode. Some electrons are turned back by the negative repeller potential, but then rejoin the main beam. Because of the resonant cavity structure, the moving electrons set up magnetic pulsations that react with each other to velocity-modulate the electron stream.

A more recent development is the Gunn diode, which is less expensive and is proving effective as a local oscillator. The Gunn diode is a negative-resistance device, operating within tuned cavities as a negative-resistance oscillator. A cross section of the Gunn diode oscillator system used in the RCA Weatherscout radar is shown in Fig. 10-6. It operates in the 9375-kHz range and is capable of automatic tuning to track magnetron drift such that the difference between magnetron and local-oscillator frequencies always equals the intermediate frequency. Without this feature, the if stage would not be within tuned range. The varactor diode is a reactive device capable of changing reactive characteristics as a function of a dc bias voltage. The dc bias originates from automatic frequency control (afc).

Automatic frequency control operates in two modes—search and track. When first in operation, it is in the search mode. In this mode, a sawtooth bias voltage is applied to the varactor diode, causing the intermediate frequency to sweep back and forth. A frequency discriminator originates a dc as a function of the difference between the local-oscillator and magnetron frequencies. When the difference frequency reaches the point that it is equal to the correct intermediate frequency, afc goes into the track mode. This is a feedback mode whereby the bias voltage institutes continuous corrections to maintain the correct intermediate frequency. In systems using a reflex klystron local oscillator, operation is the same, but in this case bias is applied to the repeller.

The rf energy is transmitted by waveguides, which are hollow, rectangular pipes in which the electromagnetic wave moves. Any surface that distinctly separates two regions of different electrical properties will tend to guide radio waves. Because the electric field of the wave is zero when parallel to the metal surface of the waveguide, and because the magnetic field is zero at the inner surface of the waveguide, all the energy of the wave is contained wholly within the waveguide. To lessen the attenuation caused by the conductive surface of a waveguide, the inner surface is sometimes gold- or silver-plated. The minimum size of waveguide that can be used to transmit a certain frequency is proportional to the wavelength of that frequency. A given waveguide will not operate below its cutoff frequency. Because of the large size required, waveguides are not used extensively below 3000 MHz. A dented waveguide is useless; a dent will detrimentally affect the standing-wave ratio. The interior of a waveguide must be kept clean. Otherwise, bits of solder or other foreign objects may cause arc-over. In addition, they will absorb power and become hot, perhaps hot enough to damage the waveguide.

The rf section of a radar system is contained in an assembly known as a *duplexer* or *duplexer-mixer*. Here, the receiver and transmitter are separated so that the transmitter pulse will not damage the receiver. Fast-acting "TR" tubes prevent rf energy from reaching the receiver. The use of ferrite rods makes it possible to reverse the polarity of a wave within the guides and thus permits the wave to move in only one direction along certain paths. The klystron or Gunn diode, acting as a local oscillator, is also a part of the duplexer, and a crystal mixer produces the if.

ANTENNA STABILIZATION

The previous discussion about terrain observation with airborne radar requires the antenna to be held level with the earth. For example, a stationary antenna will not be level during the pitch and roll of the aircraft while flying through rough air. After the aircraft is again level, the bright false pattern displayed during such movements will linger for several sweeps because of the persistence of the cathode-ray tube. As a result, the displays will be obscured and blurred. Therefore, to obtain accurate terrain displays, the antenna must be held level, or stabilized, despite variations in the attitude of the aircraft. Although the newer lightweight radars have fixed antennas, the more-expensive systems employ stabilization.

An aircraft in normal flight will roll at a much greater angle than it will pitch, for example, while flying at 150 knots and doing a 3° per second coordinated turn, the aircraft will roll 30°. The sensing element for the antenna stabilization system is a vertical gyro which generates pitch-and-roll command signals that are induced in the servo system, where they are compared with the closed-loop response from the antenna synchro transmitter. Any difference between the signals is amplified and used to stabilize the antenna accordingly. The gyro itself is a spinning wheel that is permitted to move in any direction by a gimbal. The gyro resists any force that may cause it to deviate from its spin axis. By proper design, the gyro can be kept level so that the airplane in effect rotates around it.

Command signals from the gyro come from two center-tapped potentiometers, one attached to the pitch and the other to the roll gimbal. Fig. 10-7 shows a simplified schematic of an antenna stabilization system. A 400-Hz reference voltage is impressed across these two potentiometers. When the airplane is level, the wipers rest at the center, or no-voltage, point. Motion about the pitch axis produces a movement of the pitch gimbal, moving the wiper of the pitch potentiometer. Likewise, motion about the roll axis moves the wiper of the roll potentiometer. The amount of either motion determines the magnitude and phase of voltage at the potentiometer wiper.

The pitch-and-roll voltages are amplified by the isolation amplifiers and applied to the stators of a resolver. This is a device that solves trigonometric relationships electrically. An rms ac voltage (E), when fed to a resolver, will have a value of E sin θ at the resolver output, where θ is the angle of the resolver shaft. In the application discussed, the resolver produces the trigonometric sum of the roll-and-pitch signals. The sum represents the elevation axis position which establishes a constant antenna pitch angle with respect to the horizontal angle. The output of the pitch-and-roll resolver is applied to the synchro that controls the tilt angle. The

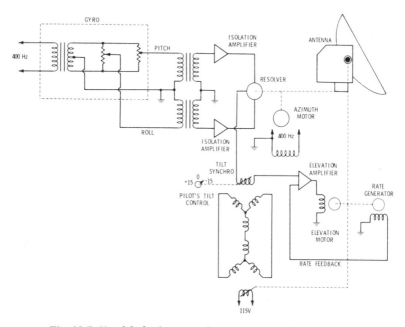

Fig. 10-7. Simplified schematic of an antenna stabilization system.

manual tilt synchro allows the pilot to set the elevation axis at a fixed angle above or below the horizontal. When the synchro transformer rotor is manually displaced, a signal voltage is induced and returned to the elevation servo-amplifier input. Here, it is added to the voltage from the pitch-and-roll resolver, and the two are applied to the tilt drive motor.

A rate generator, included as part of the elevation drive motor, produces a voltage that is proportional to the velocity of the servo motor but of opposing phase to the servo error signal. This rate voltage is added to the error signal to prevent overshoot and oscillation near the null position. For proper orientation and adjustment of a stabilization system, trim voltages are necessary. These fixed voltages bias the servo to conform to an individual installation. For simplicity, they have been omitted from Fig. 10-7.

DIGITAL RADAR SYSTEMS

Until recent years, radar systems have operated on the analog principle. That is, incoming signals were displayed and treated through analogs of time, distance, and magnitude. The simple radar concepts described in Chapter 9 are based on the analog principle. Analog display techniques deal with the signal by continuously rising or falling voltage and current levels. The scope beam is actuated by causing it to move as an analog of arriving signals and antenna position. Digital techniques, on the other hand, consider the surveillance regime as being divided into digital cells in which a target is or is not present. The ability to detect a target at any given point in space is still dependent upon the same performance parameters; that is, frequency, power, pulse width, sensitivity, etc. In the case of weather radar, however, digital techniques make it possible to present more meaningful displays by digitizing video intensity as a function of return signal strength. Moreover, large-scale integrated-circuit logic (ICL) chips make it possible to confine enormously complex circuitry into light packages and eliminate mechanical devices such as synchros and servos.

The scope display in digital radar employs a raster similar to television. The raster is stepped incrementally through the surveillance sector. The presence of a signal is indicated by its X and Y coordinates. Antenna position is interpreted by sine or cosine relationships, and the antenna may be "stepped" through the surveillance sector. Incoming signals may be stored in succession for digital interpretation. The system is controlled by time relations from a central clock.

Antenna stabilization is not included in the system to be described in the following. Because target returns are successively integrated and processed, the weather sector appears reasonably correct on the screen notwithstanding normal aircraft attitudes encountered in cruise, although more complex digital radar does employ antenna stabilization.

The RCA Weatherscout I (Fig. 10-8) is an example of an ultralight digital radar for use with single-engine aircraft. The antenna, together with the receiver/transmitter unit, may be mounted within the leading edge of a wing without disturbing aerodynamic characteristics. The complete system consists of only two units—the receiver/transmitter and the indicator.

The system operates on either 14 or 28 volts dc. Range mark figures and operating modes are displayed directly on the cathode-ray tube alphanumerically. Four signal strengths—black and levels 1, 2, and 3—are displayed. In the cyclic mode, thunderstorm returns are flashed on and off. A rectangle at the lower left of the scope screen continuously shows the video intensity of the three levels. Using large-scale integration (LSI), only three chips provide all timing, memory, and computation functions. A block diagram of the system is shown in Fig. 10-9. The system is very simple to operate, with only four controls. Range, intensity, and antenna tilt are controlled with conventional knobs, with a switch knob for the cyclic-mode feature.

The following discussion presupposes a knowledge of digital electronics on the part of the reader:

Target returns are displayed at one of four video levels—0, 1, 2, or 3. Level 0 is shown as a dark screen because of weak or no returns; levels 1, 2, and 3 are shown as progressively brighter displays which represent increasing rainfall, from least to heaviest. The magnetron generates 9345-MHz pulses of about 1-kW peak power and feeds them to the circulator. In the transmit mode the circulator channels magnetron energy to the antenna, and in the receive mode the circulator channels the return signal energy to the receiver circuits. A T/R limiter isolates the receiver circuits from transmitter energy that leaks through the circulator.

Return signals are passed by the circulator and T/R limiter to the balanced mixer. In the balanced mixer, a 9375-MHz cw signal from the voltage-controlled local oscillator is mixed with the received 9345-MHz signal. The mixer yields an intermediate frequency of 30 MHz which is fed to the if and video amplifiers. The resultant video drives a 3-level digitizer, and the 2-bit coded video output is sent to the indicator for display.

Automatic frequency control (afc) samples the frequency of the if amplifier and applies a control voltage to the voltage-controlled local oscillator. The local-oscillator frequency then changes as required to track magnetron drift.

The timing and control logic of the indicator sends a 228-Hz signal to the receiver/transmitter. This is used to trigger the modulator at

Fig. 10-8. The RCA Weatherscout I digital weather radar system.
(*Courtesy RCA Corp.*)

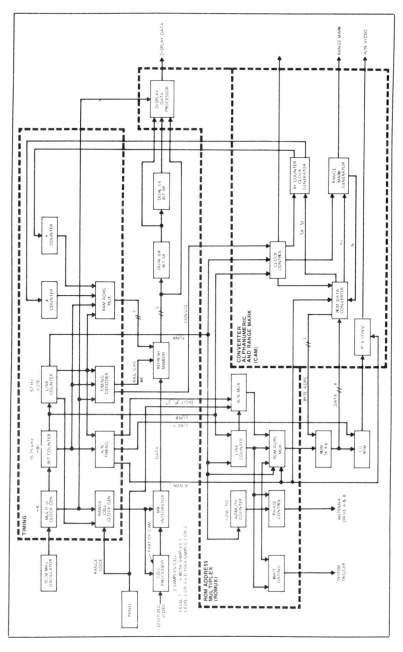

Fig. 10-9. Block diagram of the RCA Weatherscout I logic circuitry.
(*Courtesy RCA Corp.*)

a 228-Hz rate, and provide the if amplifier with an agc and a blanking gate. The blanking gate disables the receiver for 30 microseconds to keep the transmitter pulse from being displayed. It also generates the sensitivity time control (STC) function which lowers receiver sensitivity at close ranges and allows it to increase gradually at longer ranges.

The modulator, when triggered by the indicator, triggers the magnetron with high-power 4- or 10-microsecond pulses. When a 90-mile indicator is used, the transmitted pulse width is 4 microseconds on the 12-mile range. When a 120-mile indicator is used, the 4-microsecond pulse width is used on both the 12- and 30-mile ranges.

The antenna parabola is positioned in both azimuth and elevation. The azimuth-drive stepper motor drives the antenna in azimuth in approximately 0.1° increments in response to timing signals from the indicator. The azimuth-drive stepper motor steps the antenna 767 steps in one direction to scan 90°, then reverses and steps the same amount in the opposite direction. Actual dish movement requires only 45° to move the electrical beam position 90°; this is accomplished through antenna geometry. Two 57-kHz square waves are received from the indicator over two wires. Depending upon the phase relationship of these two signals, the azimuth-drive stepper motor is driven either clockwise or counterclockwise.

The timing and control logic originates timing chains from a crystal-controlled 10.08-MHz oscillator. This frequency is counted down to the 15.75-kHz line sweep rate, through the 57-Hz field refresh rate, and down to the 3.4-second period antenna scan rate. The timing and control logic also decodes and retimes the timing chains to provide timing signals such as R/T trigger, antenna step signals, sweep syncs, memory address, etc.

The range clock generator generates 128 pulses of a period scaled to match the selected range. The integrator accumulates the last four radar returns for each of the 128 range cells before the resultant data is stored to memory. The memory consists of two 16K random-access memory (RAM) chips organized into a 128 × 128 X-Y matrix with two bits assigned to each grid. It accepts data from the integrator to the location defined by the scan converter, and is read out synchronously with the sweep timings.

The scan converter generates X-Y addresses for each of the range cells at the angle defined by sine and cosine values provided by a read-only memory (ROM). The read-only memory receives the antenna angle information from the timing logic. The data readout buffer stores one column of memory data to convert a slow memory data readout rate to a faster data display rate. It also provides the cyclic function. The range mark and alphanumeric (A/N) generator generates background videos; i.e., three range marks, range mark alphanumeric labels, and mode alphanumerics.

Circuits that accomplish the cell logic are organized around three integrated-circuit chips:

Timing
ROMUX   (ROM Address Multiplex)
CAM   (Converter, Alphanumeric, and Range Mark)

Digitized data from the R/T unit is first processed in the CAM chip before it is accumulated by the integrator. Information is sampled on both edges of the buffer clock, thus twice per cell. For level-1 data, both bits must be logical 1; for level-2 or level-3 data, either bit at logical 1 is sufficient to establish that level. The algorithm for the processor is:

	First Sample			
	0	1	2	3
0	0	0	2	3
1	0	1	2	3
2	2	2	2	3
3	3	3	3	3

Second Sample (rows 0, 1, 2, 3) →Cell Processor Output

Input to and output from the cell processor is complemented gray code.

The "main-bang" (MB) integrator consists of a 256K × 4 ROM and a quad 128-bit shift register. There are two modes of operation—data acquisition and memory update. During data acquisition mode, the digitized radar echo data is accumulated cell by cell. During the memory update period, accumulated data is decoded, transferred to the appropriate location of the refresh memory, and loaded into the shift register. The pulse repetition frequency is 228 Hz and the update rate is 57 Hz, or four "main bangs" between memory updates. Integration is over the last four "main bangs" plus the result of previous integration. The integrated data for the nth update may be expressed as:

$$I_n = \text{Integer of} \left[\frac{A_n + B_n + C_n + D_n + I_{n-1} + 2}{5} \right]$$

where A_n, B_n, C_n, and D_n are the data from the last four "main bangs" with values of 0 through 3. Some exception is made in order to reduce interference from other radars.

The refresh memory has storage capacity for a 128 × 128 cell matrix, each cell containing two bits to define one of four intensity levels. It consists of two 16K × 1 dynamic random-access memories (RAM), one for each intensity bit. The 14-bit RAM address is strobed out in two sequences, seven bits per sequence. The first seven bits are the row address, strobed by RAS; the next seven bits are the column address, strobed by CAS. Read or write operation is controlled by write-enable signal WE. The row address is assigned to vertical image cells and the column address is assigned to horizontal image cells, as follows:

RAM ADDRESS	A0	A1	A2	A3	A4	A5	A6
ROW	$Y2^0$	$Y2^1$	$Y2^2$	$Y2^3$	$Y2^4$	$Y2^5$	$Y2^6$
COLUMN	$X2^0$	$X2^1$	$X2^2$	$X2^3$	$X2^4$	$X2^5$	Right

$Y2^0$ through $Y2^6$ is the vertical binary address; $X2^0$ through $X2^5$ is the horizontal binary address relative to the vertical center line. Left or right is specified by column address A6.

Additional logic operations are required to generate and convert range mark, azimuth information, and scan conversion. The scan converter converts ΔX and ΔY. A separate printed circuit board accepts sync information and digitized video, provides line and field deflection, cathode-ray tube voltage levels, and video drive to the cathode-ray tube.

The foregoing discussion describes essential logic operations provided by three chips which form the system microprocessor. The R/T unit contains the equipment which generates the pulse, receives the returning echoes, and converts the discrete echoes to digital quantities.

RADAR PERFORMANCE ANALYSIS AND MAINTENANCE

A pure sine wave has a single frequency and is of constant amplitude. Any change of the sine wave from one state to another creates other frequency components. A short wave in the form of a pulse cannot be said to be of one single frequency. Generally speaking, the shorter the pulse, the sharper the rise time will be, with a resultant increase in frequency components. In other words, the *energy* of the impulse is contained in a spectrum. This phenomenon was first revealed by the French mathematician, Fourier, long before the advent of modern electronics. The spectrum of a pulse can be represented by plotting energy *versus* frequency. The results depend on the shape of the pulse; a short, perfectly square pulse would have the energy spectrum shown in Fig. 10-10.

The greatest energy distribution is at the center, or what may be called the *carrier*, frequency. The rest of the energy is in the sidebands, and any distortion in the pulse is indicated there. Spectrum analysis is therefore an important part of radar performance analysis. The performance of a magnetron depends on the magnetic field, the load impedance, and the modulator pulse. Defects in the "plumbing" (waveguide circuitry leading to the antenna) and improper modulation pulses will tend to distort the spectrum. Load impedance and pulse amplitude also affect the operating frequency of the magnetron. *Moding* is a condition whereby the magnetron "pulses" at frequencies somewhat removed from its design frequency. It is revealed by examination of the magnetron spectrum, and is due to the same conditions that cause frequency shift.

The narrower the pulse, the wider the frequency spectrum it occupies. As a rule of thumb, the energy of a pulse is contained in a frequency band equal to twice the reciprocal of the pulse width.

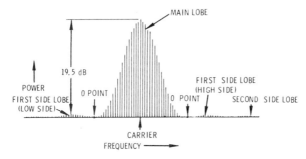

Fig. 10-10. Ideal spectrum for a rectangular pulse.

Thus, a pulse of one microsecond (10^{-6} second) will have a reciprocal of 1 MHz (10^6 Hz). The bandwidth of the receiver must therefore be at least twice that, or 2 MHz. The wider the bandwidth, the more noise energy is admitted to the receiver, however. Narrower pulses mean better definition, as previously pointed out—but more receiver noise means less sensitivity and hence range. Therefore, the noise energy at the input of a receiver always determines the sensitivity of a communications system. The minimum discernible signal is considered equal to twice the noise energy. Measurement of the noise figure is thus an important factor in radar performance analysis.

Performance analysis requires four test instruments—a spectrum analyzer, signal generator, power monitor, and frequency meter.

The IFR, Inc., Model RD-300 weather radar test set (Fig. 10-11) was designed to simplify radar testing and troubleshooting by combining several functions in a single unit. All rf tests and measurements are made with one coaxial cable connection to the radar. There is no need to reconfigure the test equipment to measure different parameters.

A signal generator system automatically acquires and tracks the magnetron frequency. This eliminates retuning the signal generator to compensate for magnetron or signal generator frequency drift.

The signal is modulated to simulate storm cell echoes in several modes including contour mode used for calibration and checkout of contour threshold circuits. Range delay of these returns is calibrated in microseconds or nautical miles. Multiple returns can be generated to check and adjust the range ring display. The auxiliary modulation mode develops narrow short-range pulses for testing multimode radars.

Fig. 10-11. The IFR, Inc., Model RD-300 weather radar test set.
(*Courtesy IFR, Inc.*)

Measurement of magnetron frequency and pulse-repetition frequency appears on the digital display. Magnetron condition and spectrum characteristics can be checked using the detector and discriminator outputs. Peak power measurements are possible with full-scale sensitivity of 12 kW peak (or kW peak with external 10-dB attenuator).

The intermediate-frequency signal generator covers from 20 to 70 MHz and can be swept. The 2 volts rms maximum output can be used for high-level if or afc testing. Bandwidth and center frequencies are measured by the marker-frequency generator during swept if tests. The RD-300 also provides an antenna simulator output which allows the display assembly to be swept at a variable rate.

Maintenance of the simple ultralight radar systems of today is not difficult. Repairs can seldom be made in the aircraft. After a trouble has been isolated to a specific unit, the unit is taken to the shop. An installed system should be inspected as follows:

1. Check radome and waveguide for damage.
2. Check units for physical damage, and firm mounting on shock mounts or base.

3. Check all cables for wear, tightness, and proper connection.
4. Check control switches and knobs for correct positioning and tightness.
5. Perform an operational check.

When operating a radar on the ground, it is important that no building, airplane, or other large metal object is within 150 feet of the antenna; the large quantities of power reflected from these objects could damage the crystals in the receiver mixer.

11

Gyroscopes and Servo Systems

In order for the airplane to become a practical vehicle, two important techniques had to follow its development. These were associated with the *gyroscope* and the *servo system*.

The technology of the servo system is the intriguing application of principles that apply to all of nature, namely, control through the observation (or sensing) of conditions that depart from a desired course of action. The importance of a servo system is that it is a true control system and can supplant human control functions. In fact, servo systems can be made superior to human control in performing many functions. The lightweight servo system has many uses, but is most important in autopilots and flight director systems.

The gyroscope makes control in three-dimensional space possible. It acts as a sensing device capable of detecting the slightest motion on a given axis, hence it may be used for directional control in pitch, roll, and yaw. The gyroscope is mandatory in the control of an airplane when it is impossible to refer to the earth.

GYROSCOPES

A French physicist, Jean Bernard Foucault (1819–1868), was among the first to investigate the principles of the gyroscope. In 1850, he used a gyroscope to demonstrate the rotation of the earth. The basis of the gyroscope is a spinning mass, usually in the form of an accurately balanced wheel.

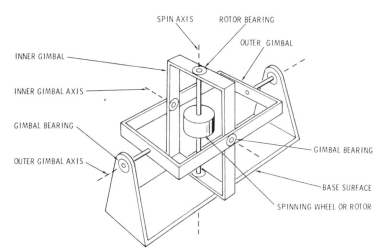

Fig. 11-1. Gyro and gimbal mounting permitting two degrees of freedom.

It is a basic law of physics that the axis of a spinning wheel will resist turning or displacement. Once a wheel is in motion and the axis of rotation established, a measurable amount of force is necessary to turn the axis. The idea is to mount a rotating wheel in such a manner that it may retain its position in space while its mounting moves around it. This unique type of mounting is known as a *gimbal,* shown in Fig. 11-1. The entire structure is called a gyroscope, derived from the Latin *gyrus* and the Greek *gyros,* meaning *a circular motion or revolution;* and the Greek *skopein,* meaning *to view.* Although the term gyroscope is technically correct, pilots and engineers employ the shorter term *gyro,* which will be used hereafter.

Notice that the structure shown is free to move in all directions. The shaft upon which the wheel rotates can be placed in any position relative to the base surface. Now, if the wheel is set in motion, the spin axis or shaft will remain fixed in space. In other words, the shaft is isolated in space from the base platform. Relative angles between the gyro axis and the base surface can now be measured.

For example, if the axis is placed in a vertical position and the base surface is mounted in an airplane, pitch and roll of the aircraft can be noted by the relative position of the spin axis. Such an arrangement is called a *vertical* gyro. If the spin axis is horizontal, the arrangement will sense direction, and is called a *directional* gyro. Both the vertical and the directional gyro are classified as "two-degrees-of-freedom" gyros. A "single-degree-of-freedom" gyro is shown in Fig. 11-2.

The latter setup demonstrates how the gyro axis resists spatial movements. For example, if the base surface were rotated about the quadrature axis (a plane parallel to the base surface), the spin axis

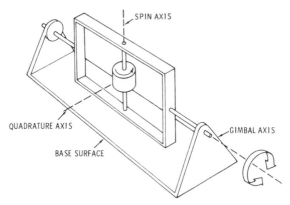

SPIN AXIS

QUADRATURE AXIS

GIMBAL AXIS

BASE SURFACE

Fig. 11-2. The "single-degree-of-freedom" mounting.

would tend to twist the gimbal. The amount of torque or twisting force is directly proportional to the rate at which the base surface is being turned. Hence, we see here a demonstration of a so-called *rate gyro.*

Two classifications of gyros have been discussed so far, those with single and double degrees of freedom. There is also a "three-degrees-of-freedom" gyro. This may be made up of two, "two-degrees-of-freedom" gyros on a suitable gimbal. Or it could be three gyros, each with a single degree of freedom, suitably gimballed. A "three-degrees-of-freedom" gyro is called a *stable platform.* It reflects angular motion of the base surface relative to a three-dimensional coordinate system. This latter requirement is seldom necessary in commercial and private aircraft, and thus will not be discussed here.

The axis of the spinning mass (wheel) of the gyro resists a change in direction because of a property known as *angular momentum.* This is the potential force present in the wheel, and is a function of rotating speed (angular velocity) and the actual mass of the wheel. As the wheel spins, each molecule of the mass is responsive only to the spindle (axis) of the structure. The molecules are "orbiting" in the same sense that a satellite orbits the earth or another planet. Gravity holds the satellite in orbit. Rigidity of the metal itself holds the molecules in the case of the wheel. Movement of the axis against its dynamic force is called *precession.* Whenever a gyro is precessed by exerting a torque force upon its axis, the amount of torque required is given by

$$T = MR \qquad (\text{Eq. } 11\text{-}1)$$

where,

T is the torque in ounce-inches or newton-meters,
M is the momentum of the wheel in ounces per square inch per second or grams per square centimeter per second,
R is the rate in degrees per minute or radians per second.

This equation is known as the *Law of Gyroscopes*. More important for this discussion is the determination of the direction in which these forces act. This can be done by using the "right-hand rule," shown in Fig. 11-3. When torque is applied in the direction indicated by the torque vector, the spin vector tries to move into the torque vector. In other words, if torque is applied at right angles to the spin axis, the gyro will respond with a torque at 90° to the applied torque.

This property, which is always surprising to the uninitiated, has often been used for humor by student technicians. Certain types of gyros with very smooth bearings run for a considerable length of

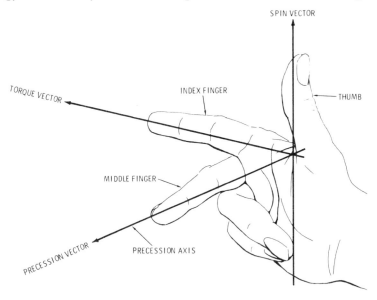

Fig. 11-3. The "right-hand rule" of the gyro.

time after the power is removed and the gimbals are locked. A gyro would be "revved up," then disconnected from the power and placed in a suitcase or other container. When an unsuspecting new student was told to move the container, a change in direction resulted in a violent reaction at right angles to the intended change.

The property that surprised the student in the latter example is used in *rate* gyros. Rate gyros are single-degree-of-freedom instruments that measure rate of turn. The first use of gyros in airplanes was as rate-of-turn indicators. Before vertical gyros were introduced, this was the only means of preventing "graveyard spirals" under conditions of bad visibility. The principle of a rate gyro is shown in Fig. 11-4.

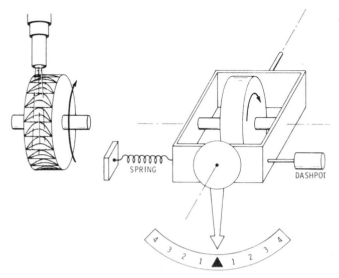

Fig. 11-4. A "single-degree-of-freedom" gyro used as a rate-of-turn instrument.

When the aircraft is on a constant heading, the gyro remains stable and the needle is centered by the spring and dashpot. When the heading is changed, torque is exerted, and the gyro returns torque perpendicular to the applied torque. More torque exerts more pressure on the spring, indicating a higher rate of turn.

Illustrated in Fig. 11-5 are two types of rate-of-turn indicators. Of the two types shown, the 2-minute indicator (Fig. 11-5A) is the older. If the instrument is accurately calibrated, a single needle-width deflection means that the aircraft is turning at 3° per second. This is a "standard-rate" turn in which the aircraft completes 360° in

(A) Two-minute turn indicator.

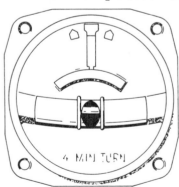

(B) Four-minute turn indicator.

Fig. 11-5. Two types of rate-of-turn indicators.

two minutes. On the 4-minute indicator (Fig. 11-5B), a single needle-width deflection shows when the aircraft is turning at $1\frac{1}{2}°$ per second, or "half standard rate" (four minutes for a 360° turn). From a comparison of the indexes on the two instruments, it can be seen why the 4-minute indicator was developed. The half standard-rate turn (half needle-width deflection on the 2-minute indicator) used by high-speed aircraft is more easily read on the 4-minute indicator.

Below the rate-of-turn needle is a ball that becomes displaced when the ailerons are not properly coordinated with the rudder. This happens when the roll forces acting on the airplane are not perpendicular to the wings. Flying on the rate-of-turn indicator can be tedious. To hold a heading, rudder pressure is applied opposite to incidental movement of the turn indicator. Aileron pressure is applied in the direction of the ball to keep the ball centered at all times.

Early flyers were very determined. When the only attitude indication came from the rate-of-turn indicator, heading and maneuvering were done in the manner just described. Considering that the pilot was also tormented by the steady tones of the "AN" radio range and possibly considerable static, one or two hours of such flight could be a demanding task.

Power for Gyros

A pioneer method for powering gyros that is still in use is by vacuum. Vacuum was developed by the use of a *venturi*, shown in cross section in Fig. 11-6. As the aircraft moves through the air, some

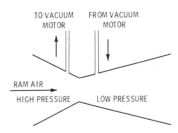

Fig. 11-6. Basic principle of the venturi.

air is scooped and compressed in the forward part of the venturi. As air passes through the narrow throat and decompresses, much less pressure appears in the space behind the throat. This is according to the important *Bernouli* principle which is intrinsic to aerial flight.

A considerable difference in air pressure is caused to exist on either side of the throat. Hoses connect to two holes on either side of the throat, allowing a powerful volume of air to pass. This air passes through a nozzle to cause the gyro to rotate through turbine action. Rotational velocities as high as 100,000 r/min can be attained.

253

Outside venturis are now used only on training aircraft. This is because of the danger of ice forming that could clog the venturi, causing loss of instruments at critical moments. Simple vacuum pumps connected to the engine are an effective modern substitute.

Electric power for gyros is highly effective. Usually the gyro is driven between 4000 and 24,000 r/min. Since angular momentum is the actuating force, gyro rotors should have as large a diameter as possible. This is accomplished by turning the gyro motor inside out, with the stator inside and the rotor rotating around it. Thus, the motor itself becomes the rotor, as shown in Fig. 11-7.

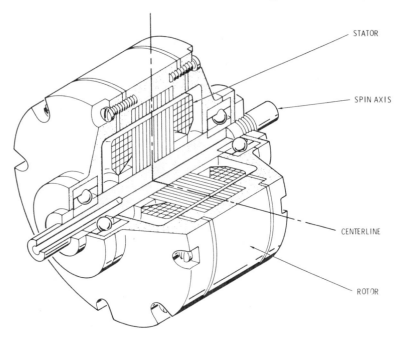

Fig. 11-7. Cross section of a gyro motor showing how the rotor surrounds the stator, opposite the manner of a conventional induction motor.

SERVO SYSTEMS

The airborne servo system has now been perfected to a respectable degree. It is usually made up of standard components which, in turn, have been subjected to severe tests to meet military requirements. Servo theory is a subject that intrigues specialists with mathematical leanings, and much has appeared in print on this subject in the past three decades. Most practical systems, however, are put together using "cut-and-try" techniques and standard components.

Before going on to servo mechanisms, it is necessary to discuss these components.

Synchro Components

The synchro is a kind of transformer with a rotating winding (see Fig. 11-8). Synchros are used to transfer angular data by wire. They were first perfected for use on weapons such as antiaircraft guns and ship guns. A synchro generator (TX in Fig. 11-9) consists of a rotor with a single winding and a stator with three windings separated electrically by 120°. The rotor is coupled to the shaft that provides the angular input data. The rotor winding is excited from an ac

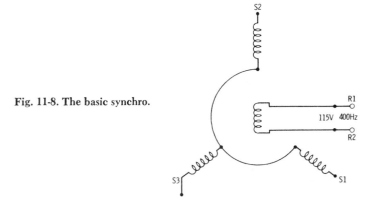

Fig. 11-8. The basic synchro.

source. Voltages that are induced into the three stator windings are a function of the rotor angle. As shown in Fig. 11-9, the synchro receiver, TR (sometimes called repeater), appears electrically identical to the synchro transmitter. Its mechanical construction varies, however, in that special antifriction precautions are taken. The rotor is free to turn, and presents the angular information at the receiving end. The rotor assumes a position resulting from the phase and magnitude relations of the voltages induced into the windings of the transmitter.

To examine the principle of the synchro, it must be remembered that a synchro is basically a transformer device. Induced voltage of the rotors is a function of the cosine of the angle of the rotor to each of the respective stator windings. There is usually a step-down ratio between rotor and stator (commonly 2.2:1) which is maintained simply for electrical convenience. No more than two of the stator voltages can be the same for a given angular position.

Synchros are not three-phase devices. Induced voltages will either be in phase or 180° out of phase with the exciting voltage. The three stator voltages plotted versus angular rotor position are shown in

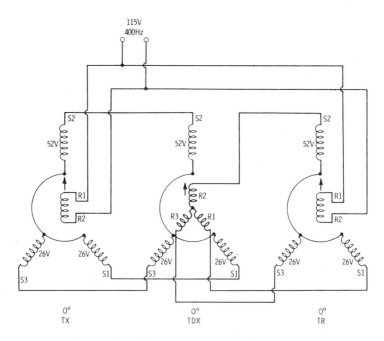

Fig. 11-9. The synchro differential (TDX), placed between a synchro transmitter (TX) and receiver (TR), will add to or subtract from the angular position of the transmitter.

Fig. 11-10. Effect of the three stator currents is to erect a resultant electromagnetic field in the synchro receiver, which will cause the rotor to assume the same position as the rotor of the transmitter. When the transmitter and receiver rotors are so aligned, they are said to be *nulled*, and the only excitation current flowing is the small

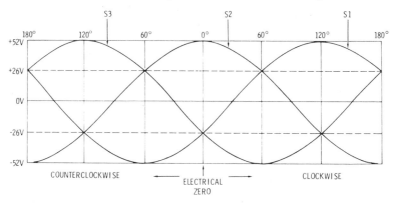

Fig. 11-10. Voltage relationships versus angular position in a synchro.

amount resulting from transformer losses. The following rules of thumb apply to relationships of the resultant field in synchros:

1. The field of each respective transmitter stator coil opposes the field of the rotor.
2. The resultant field of the transmitter stator opposes the field of the rotor.
3. The resultant field at the receiver stator opposes the resultant field at the transmitter.
4. The receiver rotor field tends to line up with the stator resultant.
5. At null, the rotor coils induce equal and opposite voltages in the respective stators, thus canceling stator current.

In Fig. 11-9, note that the windings have standard designations. The angular position of the shaft is considered to be zero when the rotor is exactly aligned with S2. A synchro may be driven to its electrical zero by the jumper method, shown in Fig. 11-11. The jumpers assure that voltages S1 and S2 are exactly equal, and the voltage of S2 is at maximum; hence, the rotor is electrically in line. With the leads thus jumpered, the mechanical dials indicating angular quantities may now be set to zero.

It is possible to change the action of synchros by changing the stator hookups. Sometimes this is done deliberately, but occasionally it happens through error while troubleshooting. Here are the effects of reversing stator windings:

S1 and S3 Reversed—Receiver shaft turns opposite to the transmitter.
S1 and S2 Reversed—Receiver rotor nulls 240° from the transmitter, and turns in the opposite direction.
S2 and S3 Reversed—Receiver rotor nulls 120° from the transmitter and turns in the opposite direction.

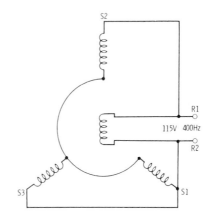

Fig. 11-11. Locating electrical zero by the shorting method.

Synchro Differential (TDX)

This device acts like a mechanical differential when placed in a synchro system. It has the same three stator windings of a synchro, but the rotor also has three windings that are similar to the stator windings. Voltages appearing at a synchro differential output rotate the resultant stator field as a function of the angular position of the differential rotor as well as the position of the transmitter rotor; thus, if θ_a is the position of the transmitter rotor and θ_b is the position of the differential rotor, the synchro receiver rotor will null at a shaft position of $\theta_a \pm \theta_b$ degrees.

A synchro differential may be either a transmitter (TDX) or receiver (TDR). As a transmitter, the shaft is held set to a selected angular position. As a receiver, the rotor is allowed to rotate to a null when energized with the input of two synchro transmitters. In this case, the differential receiver nulls at a shaft position that is equal to the algebraic sum of the two inputs.

Synchro differentials have many uses, such as in setting autopilot trim, in navigational computers, etc. Any synchro device is incapable of performing work as such. These handy, highly developed components are suitable only for transmitting angular data. When a synchro is required to apply torque, it slips from its null. Thus, inertial and frictional loads cause inaccuracy. If a synchro receiver rotor is forced from its null, the torque that must be applied steadily increases until the rotor is displaced from the field by 90°. At this point, the amount of torque decreases as the 180° position is reached. A synchro will not null at 180° if free of torque, however, since any transient displacement would cause the rotor to immediately snap back to zero.

Synchro torque is measured by *torque gradient*. This is the amount of torque required to pull the rotor 1° from null. For small angles off null, the amount of torque is assumed to increase linearly. Hence, if a given synchro has a torque gradient of 0.5 inch-ounce per degree, 1 inch-pound of torque will pull the rotor 2° off null. Torque gradient varies inversely as the impedance of a given synchro component.

Synchro Control Transformer (CT)

A synchro control transformer appears electrically identical to a synchro transmitter or receiver, but is constructed differently. The synchro control transformer produces the "error" voltage of a servo system. The stator of a CT is wound with more turns of finer wire. The cylindrical rotor also has a high-impedance winding. No voltage is applied to the rotor winding. Instead, a voltage is induced in the rotor winding whenever it is off null. Null in a CT will be that rotor position that produces no induced voltage from the resultant field of the three stator windings. The stator windings have been fed from the synchro transmitter.

In the CT, electrical zero is considered to be the position in which the rotor is at 90° to S2. In this position, there will be no induced rotor voltage. The resultant field of the stator rotates as the synchro transmitter shaft is rotated; hence, the null position rotates accordingly. There will be two rotor positions, respectively, for null and maximum. However, there will also be an in-phase and out-of-phase position. As will be explained later, it is this property that causes the servo to operate.

Basic Servo Loops

A basic servo loop is shown in Fig. 11-12. Broken down to essentials, a servo consists of an input device, an error detector, a regulator, a driving mechanism, and a load. When an error in position is sensed, a regulator apportions energy to a driving mechanism that exerts force on the load to correct the error. Energy may be in many forms such as electrical, hydraulic, or pneumatic. When electrical energy is used, the regulator becomes an amplifier driving an electric motor, and the error detector may be a control transformer with its shaft fixed to the error-sensing device.

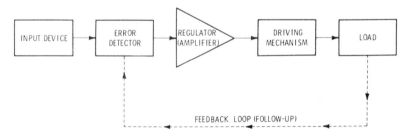

Fig. 11-12. Basic servo loop.

In the electrical servo shown in Fig. 11-13, two methods of sensing error are shown. One method uses a form of control transformer, and the other uses a potentiometer with brush attached to the error-sensing device. Both methods are based on the use of an ac reference phase to sense direction of the error; that is, whether the shaft must rotate clockwise or counterclockwise to correct the error. Energy is furnished by a two-phase induction motor. Reversing one phase with reference to the other will reverse the motor. One phase is supplied by the amplifier and is so arranged as to be 90° ahead of or behind the difference. Torque depends on energy furnished by the amplifier, which in turn depends on the magnitude and phase of

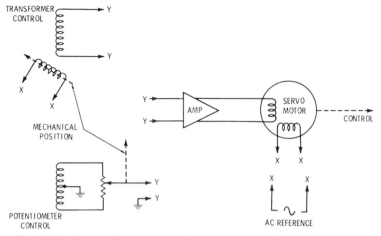

Fig. 11-13. Electrical servo system showing two methods of sensing error.

the error voltage sensed by either of the two methods. In the absence of an error voltage, one phase will be absent in the motor and motion will cease.

In the control-transformer method, the servo shaft does not move when the rotor is at 90° to the pickup coil. Phase polarity is such that rotation can be right or left; on one side of 90° the error voltage will be 180° opposite to the other, causing the direction of the motor to reverse. In the potentiometer method, the null position is at the midpoint setting of the potentiometer. On either side of midpoint, the phase will be opposite and cause the shaft to rotate in one direction or the other.

The open-loop servos just described are of a type that would be used for an autopilot, as will be described in the next chapter. Another way to use a servo is to cause a shaft to be moved a fixed amount by command. Such a method is shown in Fig. 11-14. Here,

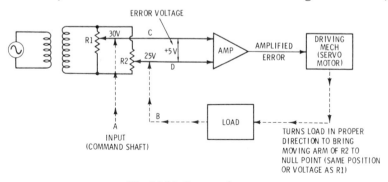

Fig. 11-14. Command servo.

the error is sensed by a potentiometer. Shaft A is the command shaft. Shaft B is mechanically connected to the servo motor. The device will null when no potential exists between C and D. Relative phase of the error voltage depends on whether the command arm is above or below the error-sensing arm; hence, motion can be so phased as to approach null.

The command servo has been highly refined, as shown in Fig. 11-15. Here the standard synchro devices previously described may be compounded to form versatile systems. Note that in schematic form, a single solid line represents the several wires that connect the

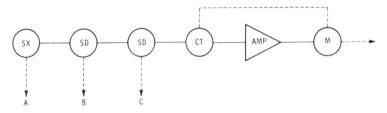

Fig. 11-15. Servo computer.

synchro devices and no reference voltage is shown; the latter is assumed at appropriate points. Mechanical shafts are denoted by dotted lines. This arrangement is actually an electromechanical analog computer. The sum of rotation between shafts A, B, and C appears on the output servo shaft. An arrangement such as this is commonly found in instrumentation (for example, the radio-magnetic indicator shown in Chapter 7, Fig. 7-16), guidance, and computer systems. Sometimes the light servos are employed solely to drive instrument pointers with very small errors.

Servo Parameters

A servo system can be rated in terms of the fidelity by which the controlled device reacts. This takes into account both output accuracy and stability. When a system is stable, output corresponds to input and there is no hunting or oscillation. Stability and accuracy are achieved in various ways, depending on performance requirements.

Stability is one problem encountered in servomechanism design. The controlled device may overshoot the ordered position due to momentum. This causes feedback from the follow-up device to become larger than the original input command signal. The error then changes polarity and applies a braking action to the controlling device. The reversed-polarity signal causes the controlling device to drive the controlled device beyond the new ordered position and the process repeats, causing overshooting or oscillation. Such electromechanical oscillations in the servomechanism are analogous to those of an electronic oscillator, where tank-circuit losses are counteracted by regenerative feedback.

In the servomechanism, the controlling device, which is a power source, is analogous to the generator, while friction losses in the servo may be likened to resistance losses of the electrical circuit. Oscillations are sustained by a repetitive supply of energy from the generator to the tank circuit, just as oscillations are sustained in a servo system by reinforcement from the controlling device.

Time delay in the controlling device may act to reinforce output-shaft oscillations and also cause continuous hunting. This comes about when a delay exists between the time an error appears and the time torque is applied to correct the error. Assume that there is no time delay in a system and the input shaft is rotated instantaneously from zero velocity to a constant angular velocity, as shown in Fig. 11-16A. Curve A shows an undamped condition in a system with no friction, and curves B, C, and D represent the response with three values of damping applied. Since no controlling time delay is involved, load oscillations are damped out in each case. However, there is a difference in the time required for the load to reach the ordered position, which is the point at which error-voltage cancellation takes place. Curve B represents an underdamped condition in which overshoot is pronounced, but oscillations finally die out at time t_3. When the system is overdamped (curve D), the load is slowed down more than that required to prevent overshoot. Thus, the speed of response of the system is materially reduced, and a longer time period is necessary for error-voltage cancellation. In a critically damped system (curve C), the value of damping is the least amount which will eliminate overshoot. Consequently, the error voltage reaches the zero value in the shortest possible time and is held at that value.

Fig. 11-16B shows the relationship between the steady-state error and angular displacement. Angular displacement increases linearly with respect to time. Although the average value of the steady-state error is zero, the system oscillates with an amplitude proportional to input velocity. Curve B, the underdamped condition, shows that overshoot is pronounced and the steady-state error is reached after an extended transient period. Thus, there is a difference in the time required for the load to reach the ordered velocity which is the point where the error-voltage cancellation occurs. In the critically damped system (curve C), the value of damping of additional friction applied is the least amount necessary to eliminate overshoot; consequently, the steady-state error is reached in the shortest time possible. However, velocity lag increases as does the steady-state error. When the system is overdamped (curve D), the load is slowed down

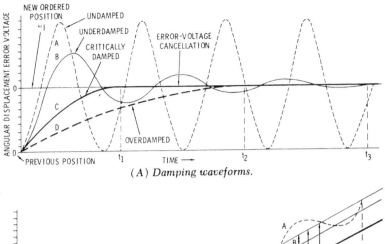

(A) Damping waveforms.

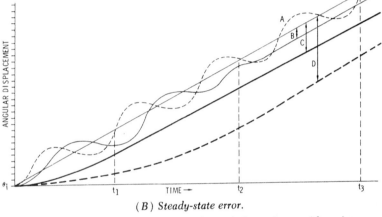

(B) Steady-state error.

Fig. 11-16. Angular displacement for theoretical step input with various
amounts of damping in a servo loop.

more than the required amount to prevent overshoot, and the speed
of response of the system is materially reduced. Thus, a longer time
period (velocity lag) is necessary to reach steady-state error.

Fig. 11-17 shows a reproduction of the underdamped curve of
Fig. 11-16. Since the assumption has been made that there is no time
delay in the system, the torque is proportional to the present error.
Here, each time the load swings past the ordered position, a reverse
torque is immediately applied to return the load to the balance point.
In this instance, there is no regenerative compensation for frictional
losses, and the oscillations are finally damped out. Relative sizes of
the arrows indicate that torque is proportional to the angular displacement.

The existence of time delay in any positioning system causes the
applied torque to be proportional to a past error, rather than a pres-

ent one. In the case illustrated in Fig. 11-17, note that torque reverses each time error reverses and maximum torque is developed at the instant the peak of the waveform, or maximum angular displacement, is reached. But when a time delay is introduced in the controlling device (Fig. 11-18), the corrective torque lags by some

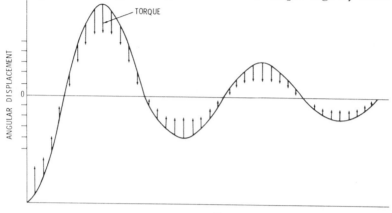

Fig. 11-17. Damping, with torque a direct function of error.

angle, and is thereby applied in the wrong direction for a short period of time after the load swings past the zero-error position. The torque arrows in Fig. 11-18 show that maximum torque is no longer developed at the peak of the waveform, but at some distance beyond the peak, this distance being equal to the lag angle. Thus, the time delay in the controlling device creates a regenerative action, reinforcing the oscillations each time the ordered position is passed.

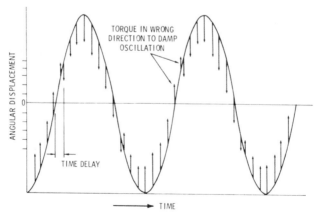

Fig. 11-18. Oscillation in the case where torque is proportional to past error.

A servomechanism can be stabilized (critically damped) by adding a factor that causes the controlling device to anticipate approach to the ordered position. This factor may be introduced by either mechanical or electrical methods.

Mechanical Stabilization—Viscous damping may be produced by a torque proportional to the output shaft speed and directly opposite to the output speed. One method is to use a dry-friction device consisting of a flywheel clamped to the servomotor shaft. The flywheel is free to rotate about a bushing, but is restrained by an oil-soaked felt washer located between the flywheel and a pressure disk. During acceleration, a viscous drag is placed upon the shaft by inertia of the flywheel. When the flywheel reaches the same speed as the motor, drag is reduced to zero.

Another method of damping is to use an electromagnetic damper. This device consists of a high-inertia permanent magnet that is free to rotate, a thin copper cup rigidly fastened to the servomotor shaft and located in the magnetic field of the permanent magnet, and a soft-iron keeper fixed to the frame of the servomotor that completes the magnetic circuit. When the servomotor accelerates, a current is induced in the copper cup, producing a magnetic field that opposes the field of the permanent magnet. The resulting torque produced causes the high-inertia permanent magnet to rotate, and when this magnet attains the same speed as the motor shaft, damping action is minimum. When the servomotor decelerates, damping action is reversed.

Both inertia and electromagnetic-friction dampers act during acceleration and deceleration of the servomotor and have little damping effect during the steady-state condition.

Electrical Stabilization—Damping output-shaft oscillations can be accomplished electrically by compensating for the time delay in the controlling device, by reducing the gain (output) of the servo amplifier, or by integrating the error signal. To decrease time delay in the controlling device, a signal that is proportional to the time-rate-of-change of the input/output position error is added to the error voltage. This method, shown graphically in Fig. 11-19, is called *error-rate damping*. An advantage of error-rate damping is that when the error-rate signal is added to the error, the servomotor accelerates more rapidly because error rate equals the velocity of the input signal. This results in a decrease in velocity lag caused by inertia, but not the velocity lag caused by friction. However, the steady-state error is reduced.

Another method of obtaining electrical damping is to use a small generator, or tachometer, coupled to the output shaft to produce a velocity (derivative) voltage proportional to the rotor speed. This voltage, having a polarity opposite to the error voltage, is combined with the error voltage and applied to the controlling device. When the error signal decreases as the ordered position is approached, the velocity signal exceeds the error and causes a braking torque to be applied to the motor. The tachometer voltage then decreases as the motor stops. Thus, the rate-of-change of the output increases the damping, since the output torque, instead of being proportional to the output position alone, is also a function of the time-rate-of-change (change of velocity) of the output position. This method of damping is called *output-rate damping*.

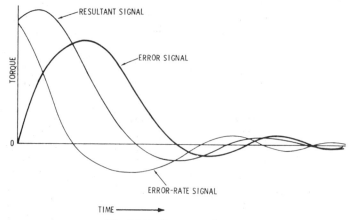

Fig. 11-19. Error-rate damping.

A third method of electrical damping, called *integral control*, depends on integration of the error signal. Integration of the error signal is the summation of the error in successive time intervals so that as the time intervals are made smaller and more numerous, the summation becomes equal to the time integral of the error, hence the name integral control. By integrating the error, an effect of increased gain occurs as the output position approaches the input. Also, the integral of the error, when added to the error signal, increases the total error and drives the output closer to the input position. After a period of time, voltage caused by integrating the error will reduce the steady-state error practically to zero.

Integral control is generally used in addition to the viscous-damped and error-rated damped systems. Viscous damping increases the steady-state error while overcoming the transient errors, and error-rate damping improves the transient response without noticeable improvement in reduction of the steady-state error. The error-integral-control damping method reduces the steady-state error of the system and also increases the effective gain and torque of the controlling device.

THEORETICAL ANALYSIS OF A SERVO LOOP

All human activity that involves controlled motion can be considered as a servo loop. A car being driven is a closed loop, with the driver sensing errors and applying pressure to steering wheel, brake, and accelerator. In an airplane, the pilot observes the horizon and his instruments, thus exerting control. A major difference between the control of an airplane over that of a car is that an airplane is capable of motion in three dimensions. We can say that where a car is only capable of yawing motion, the airplane has freedom of motion in yaw, pitch, and roll. This will be discussed in the next chapter. However, at this point we will develop the equations of a closed loop. These equations will reveal the relationships for the three important damping conditions—oscillation, critical damping, and over-damping.

We can infer intuitively that all control motion is angular, or at least may be expressed as angular motion. That is, control is exerted by moving something on an axis, for example, a joystick, control wheel, throttle quadrant, etc. Motion, then, involves torque exerted over a period of time. Resistance to the torque will be in the form of inertial force (mass), damping (friction), and stiffness. The latter term, stiffness, refers to any physical property that prevents the mass from moving as a function of torque versus time. It may be in the form of shaft springiness, human reflex, etc. Under these conditions the torque equation is

$$J \frac{d^2\theta}{dt^2} + D \frac{d\theta}{dt} + S\theta = 0 \qquad (\text{Eq. 11-2})$$

where,

J is system inertia,
D is damping,
S is stiffness,
θ is angle of motion.

The equation is set equal to zero, thus, steady-state displacement is zero and the entire motion can be considered as a transient. The characteristic equation for Eq. 11-2 is then written in the operational form

$$P^2 + \frac{D}{J} P + \frac{S}{J} = 0 \qquad (\text{Eq. 11-3})$$

Both Eqs. 11-2 and 11-3 are in the same form as a system capable of oscillation. It is known from physics that the natural frequency of oscillation, F, of an oscillatory system is

$$F = \frac{1}{2\pi} \sqrt{\frac{S}{J}} \qquad (\text{Eq. 11-4})$$

Also, the damping factor, h, of a mechanical system is

$$h = \frac{D}{2\sqrt{SJ}} \qquad \text{(Eq. 11-5)}$$

Substituting Eqs. 11-4 and 11-5 in 11-3 results in

$$P^2 + 2h2\pi FP + (2\pi F)^2 = 0 \qquad \text{(Eq. 11-6)}$$

This is a quadratic equation and it will have two roots. Solving through use of the quadratic formula results in

$$P_1 = -(h - \sqrt{h^2 - 1})2\pi F \qquad \text{(Eq. 11-7)}$$

$$P_2 = -(h + \sqrt{h^2 - 1})2\pi F \qquad \text{(Eq. 11-8)}$$

For any given moment, t, the angle of the system is

$$\theta = \theta^1 + \theta_1 \exp(P_1{}^t) + \theta_2 \exp(P_2{}^t) \qquad \text{(Eq. 11-9)}$$

It is possible to define three conditions as a solution to the foregoing. These are $h > 1$, $h < 1$, and $h = 1$. Solving for the condition $h > 1$

$$\theta(t) = \frac{\theta_0}{2\sqrt{h^2 - 1}}[h + \sqrt{h^2 - 1}]\exp\{-[h - \sqrt{h^2 - 1}]2\pi FT\}$$
$$- [h - \sqrt{h^2 - 1}]\exp\{-[h + \sqrt{h^2 - 1}]\}$$
$$\text{(Eq. 11-10)}$$

Solving for the condition $h < 1$

$$\theta(t) = \frac{\theta_0}{\sqrt{1 - \epsilon^2}}\epsilon^{-h\omega_0 t \sin[\sqrt{1-h^2}]}\omega_0 t + \phi \qquad \text{(Eq. 11-11)}$$

Wherein,

$$\phi = \text{arc tan} \frac{\sqrt{1 - \epsilon^2}}{\epsilon_j}$$

When $h = 1$ (critical damping), the solution is

$$\theta(t) = \theta_0(1 + \omega_0 t)\epsilon^{-\omega_0 t} \qquad \text{(Eq. 11-12)}$$

The three solutions, $h < 1$ (oscillation), $h = 1$ (critical damping), and $h > 1$ (over-damping), are plotted in Fig. 11-20. This response curve is plotted as a ratio of instantaneous to initial angular positions as a function of time expressed in multiples of angular velocity. Notice that Fig. 11-20 resembles Fig. 11-16A.

Equation 11-12 is related to the design of aircraft control systems and has many applications. An autopilot, for example, must operate at all times with $h = 1$ or $h > 1$. As will be seen in the next chapter, a flight path or flight condition must always be approached in an asymptotic manner, in which $h \approx 1$ but not $h < 1$. The damping

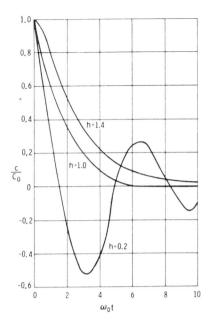

Fig. 11-20. Graph shows oscillation, critical damping, and over-damping.

factor treated in the foregoing equation is in simple form, such as friction, viscous damping, etc. Rate damping introduces an additional complex factor.

THE SLAVED GYRO

The gyro (as a sensing device) and the servo loop are employed together in several aircraft control systems. One example of the combination is in the *slaved gyro*. This is a gyro that is oriented by "slaving" it to a compass. The oldest and simplest aircraft compass is the "spirit" compass. This consists of a magnetized circular card floating in nonfreezing spirits. The float has a tendency to bob and oscillate in rough air, and its response to a heading change is slow, with considerable overshoot. Such compasses are used today only for emergency backup. A next step in compass improvement is to mount a housing containing a floating magnet in an area remote from the cockpit. The magnetic device is connected to the cockpit by a synchro transmitter and receiver. A compass card mounted on the instrument panel displays heading. This latter type of compass has the advantage of producing an easily readable heading display, but more important, the magnetic directional pickup is mounted in some area in the aircraft which is away from magnetic influence.

The slaved-gyro compass, however, is the type used at present on all aircraft flying IFR regularly. The slaved-gyro compass consists of a directional gyro and a *flux valve,* both forming a system through coupling by a servo loop. The directional gyro previously described in this chapter has the advantage of instantly following heading changes without bobbing and overshoot. The disadvantage, however, is that a free gyro such as this will drift as much as 10° per hour and must constantly be reset. In the slaved-gyro compass, the flux valve is the magnetic sensor that maintains the directional gyro on its northerly setting. Cockpit workload is thus decreased and accurate headings assured.

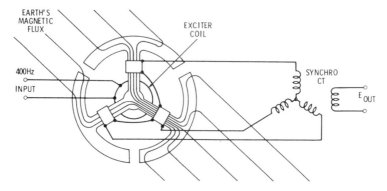

Fig. 11-21. The flux valve.

The ingenious flux-valve magnetic sensor (Fig. 11-21) consists essentially of a primary coil and three secondary coils wound on a metal core. The core resembles a three-spoke wheel slit through the rim between the spokes. The primary coil fits at the hub and the secondary coils are mounted on each spoke.

The primary coil is excited by an ac voltage. As current builds up on the positive half-cycle, an expanding magnetic field repels the magnetic field of the earth from the spokes, and current is induced in each coil. As the positive half-cycle current decays, the earth's flux is drawn into the spokes, again cutting the pickup coils and inducing a current of opposite polarity. Thus, a half-cycle of alternating current produces one full cycle of induced current. The negative half-cycle of exciting current duplicates the action of the positive half-cycle. Frequency of the induced voltage will be double that of the exciting current. The magnitude of the induced current in the pickup coils is proportional to the number of magnetic lines being cut. The number of lines cut varies according to the position of the spoke relative to the magnetic field of the earth. Each coil of the flux valve produces a voltage that is transmitted to corresponding stator coils of a flux-valve synchro. The transmitted voltages produce a resultant magnetic-flux vector whose position depends on orientation of the spokes relative to the earth's magnetic flux. When the rotor of the flux-valve synchro is perpendicular to the flux vector, no voltage is induced. Thus, the flux-valve synchro detects deviation from magnetic north of the earth. The output voltage can therefore be used to drive a servo.

12

Flight Directors and Flight Simulators

The preceding chapter pointed out that an airplane in controlled flight may be described as a closed loop or servo system. In a *flight director*, the pilot is a part of the closed loop, but the system continuously explores the relationship between actual aircraft position and desired position, then displays for the pilot the correct aircraft attitude necessary to reach the desired position without overrun.

The airplane in flight is a carefully controlled body that is acted upon by two positive forces, thrust and life. Part of the thrust goes toward providing flight momentum and another part overcomes drag which becomes either lifting force or is wasted in air resistance. The airplane designer analyzes these forces and forms a design based upon optimum considerations. It is also possible to analyze the forces of flight and design systems that reproduce the characteristics of an airplane. This is done by using mathematical models of a given airplane, and, with a computer, building a device that simulates an actual flight of the aircraft. Such a device is known as a *flight simulator*.

AERODYNAMIC PRINCIPLES

An airplane in smooth, continuous flight is in a state of equilibrium, with the wings developing exactly the amount of lift necessary to overcome the entire weight of the flying body. If the wing develops more lift than airplane weight, the aircraft will climb and vice versa. When flying a level course, the *center of lift* of the wing will be at a point above the *center of gravity*. "Trimming" an airplane in flight is the process of adjusting elevator and rudder surfaces to maintain a given flight attitude.

A turn must be perfectly coordinated. In the first Wright airplanes, roll and yaw were controlled by warping the wings. Such methods do not result in smooth turns, however. Just enough bank must be maintained in a turn to compensate for the outward force of the turn. Unless the amount is exact, there will be skidding and loss of altitude. The amount of bank is a tangential function of speed. The sharper banks of the jet aircraft, which can be noticed around airports, are due to the higher speed of the jets. Because the turns are perfectly coordinated, passengers notice no difference from slower aircraft.

All analysis of bodies in motion begins with the concept of a center of gravity and center of mass. It is found that no matter how a body is oriented, there will always be a common point through which gravity acts. The weight of a body may be treated as a single force that is applied at the center of gravity. More strictly, we can say that the line of action of weight always passes through the center of gravity. The center of gravity is also the *center of mass*. It can be shown that in a free body all motion takes place around the center of gravity (or mass).

Referring to Fig. 12-1, note that the center of gravity is directly over the center of lift. It is also in line with the center of thrust. This airplane is therefore in straight and level flight. The conditions for this are:

1. The algebraic sum of the horizontal forces (thrust and drag) is zero.
2. The algebraic sum of the vertical forces (lift and gravity) is zero.

An airplane that is in perfect trim and flying in still air will thus fly "hands off," which is a pleasant, although rare, experience. However, there must be no force to disturb the condition of equilibrium. Thus, an additional consideration is necessary for straight and level flight:

3. The algebraic sum of the *moments* about the center of gravity must be equal to zero.

A *moment* is the difference between two forces acting in different directions from different points in such a way as to cause rotation. In an airplane, the effect of a moment is to cause rotation about the center of gravity. There are many such forces. Air turbulence is a random, unpredictable force. An airplane can get out of control when lifting force is lost due to loss of flying speed. Nevertheless, in all

273

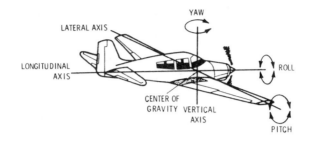

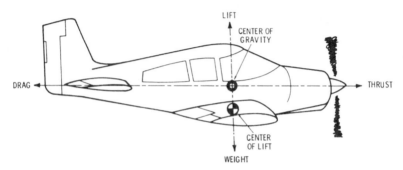

Fig. 12-1. Forces on an airplane in flight.

cases, regardless of what happens the airplane will change direction about the center of gravity. A spin is violent rotation about the center of gravity caused by a combination of inertial forces and misapplication of controls. In airplanes designed especially for acrobatic flight, the pilot is seated as closely as possible over the center of gravity so that he may precisely sense motion.

The tail surfaces of an airplane exist only for control. The rudder and elevators exert moments of force to cause the airplane to pitch or yaw. This introduces the concept of center of mass. It has been mentioned that the center of gravity is also the center of mass. The momentum of an airplane will cause the force acting on the mass to exceed the force of gravity. That is to say, the airplane will have a force acting upon it that is greater than its weight. This happens when an airplane changes direction. If only the rudder was used for a turn, the force acting on the mass would cause a skidding, mushy turn. This can be prevented by retaining the alignment of the lifting force with the force acting on the center of mass due to the change of direction. It is done by introducing roll; that is, the airplane is banked the correct amount.

The roll axis always bears a fixed relation with the wings, and if sufficient flying speed is available to provide lift, the lifting force will be in the direction of roll. Ailerons control roll, and a turn may be made by tilting the airplane with the ailerons to cause the lifting force to draw the airplane into a turn. The rudder is only used to cause the tail to point in the direction of turn. The question may be asked, "Which makes the airplane turn, aileron or rudder?" The lift of the wings in banking contributes most of the turning force, although the rudder can initiate a turn by disturbing the yaw axis, and the contribution of the rudder depends upon the design of the airplane and power setting (thrust). Actually, proper control requires well coordinated use of both rudder and ailerons. Good turns at slow rates can be made only with the ailerons, however.

During landing, takeoff, and when flying from point to point, the airplane is always operated with the center of lift and center of thrust coinciding with the center of gravity, or nearly thereto in climbs and turns. Skilled aerobatic and military pilots, using their special airplanes, however, are able to cause rotation about the center of gravity in many ways. This requires clever application of the controls.

Students of aerodynamics are concerned with many factors under such terminology as *aspect ratio, angle of attack, chord,* and so on. This is not necessary for our discussion here, however, because it is only the ultimate forces acting on an airplane that lead to the equations of control and the operation of control systems.

THEORY OF FLIGHT CONTROL

We have seen in the last chapter that for a system to be nonoscillatory, it must operate below critical damping. When this is done, all corrections will be made so that the new position is approached asymptotically. This applies whether the system is under human control or machine control. In a completely controlled flight, an airplane must:

1. Be in stable flight.
2. Occupy a desired known point in space at all times.

To meet the requirement for stability, the turn must be coordinated so that roll (bank) angle is exactly correct for rate of turn. Pitch attitude must be such that lift/drag ratio is optimum. With optimum lift/drag the power setting must be such that altitude will be evenly held. After all of the stability criteria are continuously met, the airplane must also be maneuvered to maintain the desired position in space. The position in space will have a lateral and vertical component, both measured in distance relative to the earth. The vertical component is expressed as altitude, whereas the lateral component may be in relation to a VOR station, a checkpoint, runway

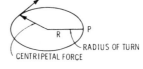

(A) *Turning about a point.* (B) *Forces acting on turning airplane.* (C) *Centripetal force.*

Fig. 12-2. Coordinated turn of airplane.

touchdown point, etc. For the discussion, we will consider only the lateral control function. This will be in the yaw and roll axes.

A basic concept in lateral control is the coordinated turn; that is, a turn in which all forces are in correct balance and there is no resulting "skid." An airplane in a coordinated "turn on a point" will follow the path shown in Fig. 12-2A. This is often used as a practice maneuver, and if correctly done the wings will be in line with the point throughout the turn.

Forces acting on the turning airplane are shown in Fig. 12-2B. The magnitude of the centripetal force is

$$F = \frac{WV^2}{R} \qquad \text{(Eq. 12-1)}$$

where,

F is magnitude of centripetal force,
W is weight in pounds,
V is speed in knots,
R is radius of turn.

The vertical component is equal to lift and the horizontal component is F, the centripetal force. From the vector diagram, Fig. 12-2B, the correct angle of bank for a coordinated turn, ϕ, will be

$$\tan \phi = \frac{\text{lift}}{\text{centripetal force}} \qquad \text{(Eq. 12-2)}$$

The lifting force is

$$F_L = Wg \qquad \text{(Eq. 12-3)}$$

where,

F_L is lifting force,
g is gravitational constant (32 ft/s^2),
W is weight in pounds.

Substituting in Eq. 12-2 results in

$$\tan \phi = \frac{WV^2/R}{Wg} = \frac{V^2}{Rg} \qquad \text{(Eq. 12-4)}$$

From this the following relationship is established:

In a coordinated turn the angle of roll is proportional to the square of the speed and inversely proportional to the radius of the turn.

Fig. 12-3 depicts the servo loop controlling the roll/yaw axes of flight. There are three input parameters necessary to meet the previously mentioned definition of controlled flight (considering only the roll/yaw axes). These inputs will satisfy the (simplified) differential equation

$$h + \phi + K\theta = 0 \qquad \text{(Eq. 12-5)}$$

where,

h is the heading necessary for asymptotic approach to the correct path,
ϕ is the roll angle from the resultant instantaneous heading,
K is the system constant,
θ is the deviation from the correct flight path.

It should be noted that θ is directly revealed by deviation of the VOR or localizer needle and ϕ is directly related to rate gyro response. Heading is sensed by the gyro compass.

In Fig. 12-4, consider that the airplane is being guided so as to reach point P at distance R. The desired flight path is PC. Distance XY is some arbitrary reference line. The aircraft is at a position offset from PC at a distance P_o from the reference line. An error signal would be equal to

$$e = -h - \phi - K\theta \qquad \text{(Eq. 12-6)}$$

This is to say that e should act in a direction to satisfy Eq. 12-5.

It is possible to design an instrument readout that would display the results of Eq. 12-6 to the pilot. By following this instrument, rather than the regular localizer needle, a pilot will now be able to make asymptotic returns to the correct flight path. This is the principle of the flight director.

The value K is a damping ratio that depends upon the aircraft weight and approach speeds. An optimum value is approximately

$$K = \frac{gR^2}{2V^2} \approx 0.7$$

when a system is used for landing approach.

The most critical lateral control problem is localizer approach to landing. Ideally, a system should have an accuracy sufficient to guide the aircraft to within 20 or 30 feet of the runway centerline. With reference to Fig. 12-4, if PC is the centerline of the runway, the error, P_e, from the runway is

$$P_e = R \sin \theta$$

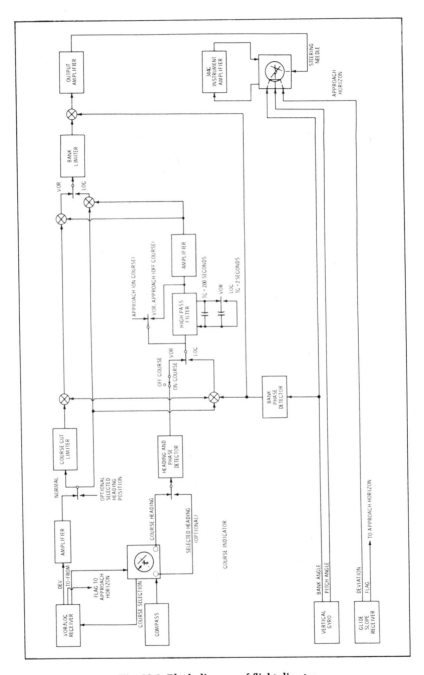

Fig. 12-3. Block diagram of flight director.

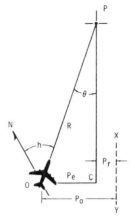

Fig. 12-4. Calculating error signal.

But we are dealing with small values of θ in all cases; therefore, $\sin \theta = \theta$, and

$$P_e = R\theta = P_o - P_r \qquad (\text{Eq. 12-7})$$

This will be one of the three control parameters, also

$$\theta = \frac{P_o - P_r}{R} \qquad (\text{Eq. 12-8})$$

The distance covered during the approach is

$$D = Vt$$

where,
   D is distance in feet,
   V is aircraft ground speed,
   t is time interval.

The component of aircraft distance normal from runway centerline in time increment dt is

$$dP_o = V \sin h\, dt$$

Again, the heading corrections are small in angle, $\sin h\,dt \approx h\,dt$, therefore the preceding becomes

$$\left(\frac{1}{V}\right)\frac{dP_o}{dt} = h \qquad (\text{Eq. 12-9})$$

This will be another one of the three control parameters.

It should be noted that aircraft heading change is inversely proportional to radius of turn and directly proportional to speed. For simplicity, we will assume that there is no wind component; hence, aircraft motion relative to air and ground are equal. In this case

$$\frac{dh}{dt} = \frac{V}{R}$$

Substituting the above in Eq. 12-4 gives

$$\frac{V\,dh}{g\,dt} = \phi \qquad\qquad (\text{Eq. 12-10})$$

Differentiating both sides of Eq. 12-9 results in

$$d^2\frac{P_o}{dt} = V\frac{dh}{dt}$$

Substituting this in Eq. 12-10, roll angle becomes

$$\phi = \frac{1}{g}d^2\frac{dP_o}{dt} \qquad\qquad (\text{Eq. 12-11})$$

We now have three control parameters. The localizer beam furnishes θ, a heading gyro gives direction, and a rate or roll gyro senses the roll correction.

The complete equation for the lateral control system of a given airplane may be stated as

$$H_1(s)\frac{1}{g}\frac{d^2P_o}{dt^2} + H_2(s)\frac{1}{V}\frac{dP_o}{dt} + H_3(s)\frac{P_o - P_r}{R} = 0$$
$$(\text{Eq. 12-12})$$

The parameters of Eq. 12-6 and Eq. 12-8 together with the measured value of θ taken from the localizer receiver satisfy the control equation.

The foregoing is a much simplified approach. As the speed of an aircraft increases, parameters become more critical. In any case, we have established a means of deriving a signal that will place an aircraft on a nonoscillatory asymptotic course to a desired path. Altitude, the vertical component, can be dealt with in the same manner.

If the pilot is included in the control loop, a visual means of presenting the steering information is required. If complete mechanical control is desired, a signal is developed to direct the servo system. The design of flight directors and autopilots requires that the equation of the specific aircraft be derived. A number of methods have been developed. The digital computer greatly facilitates autopilot and flight director design.

FLIGHT DIRECTORS

A flight director is a system that presents to the pilot a complete visual interpretation of the situation in such a manner that he needs only to move the control in a positive manner. The pilot is removed from the necessity of solving a complex kinetic equation with his own reflexes. A pilot's expression for poor control, hence, an oscillatory condition, is "chasing the needle." Training and skill are necessary to prevent this condition.

The flight director as a system does more than present steering information for asymptotic corrections of the flight path. It presents to the pilot the complete picture of the aircraft position in space relative to either runway, VOR station, ADF station, or any other navigational point. Fig. 12-5 shows the two basic instruments of a Collins Flight Director that have had long use in high-performance aircraft.

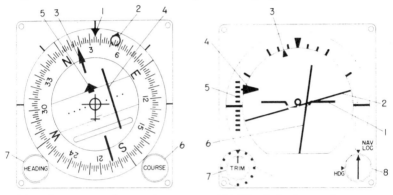

(A) *The course indicator shows a plan view of aircraft position.*

(B) *The flight director horizon is a forward view of aircraft position.*

Fig. 12-5. Two basic instruments of a Collins Flight Director.
(*Courtesy Collins Div., Rockwell International*)

The Model 331A-4 "course indicator" is shown in Fig. 12-5A. It is a "plan view" indication; that is, it displays heading angles relative to a VOR station as well as aircraft heading and deviation from a desired course. The course indicator includes the following: (1) compass heading, (2) selected heading, (3) selected localizer or VOR course, (4) deviation from localizer or VOR course, (5) "To-From" status of the VOR, (6) the course selector knob, (7) heading selector knob. The course selector knob (6) and heading selector knob (7) are used to set the respective values according to flight plan.

The Model 329B-4 Flight Director Horizon (Fig. 12-5B) provides displays of pitch, roll, and glide slope displacement. Whereas the course indicator is a plan view of aircraft position, the flight director horizon is a "forward view." The instrument is a pictorial display with the aircraft symbol (1) as the center of reference. The aircraft symbol is referred to the horizon bar (2), for pitch. Roll attitude is revealed by position of the horizon bar and bank angle indicator

(3). Relative position of the glide slope is shown by the glide slope pointer (4) and glide slope scale (5). Steering information is presented on the steering pointer (6). A manual pitch trim knob (7) is used to adjust pitch indication to allow for aircraft trim, which varies as the relationship between center of gravity and center of lift changes for any reason. The selector knob (8) determines the lateral guidance mode. The NAV LOC position refers the needle to either a localizer, VOR, or ADF radial. The HDG position is used when it is only desired to hold a heading.

When flying the correct course for a landing approach, all indicators are centered. If, for example, the glide slope indicates that the course is below the glide slope (glide slope pointer above center), an increase in pitch attitude to match the pitch bar with the glide slope pointer will bring the aircraft back on the glide path.

A great number of landing approaches have been made using flight directors with excellent results. In a large number of landing approaches, the airplane has been positioned at all times to within 25 feet of the runway centerline, notwithstanding crosswind conditions. Likewise, the glide slope approach from the middle marker has been nearly perfect.

A functional block diagram of the Collins system is shown in Fig. 12-3. Basic sensing inputs are:

VOR or localizer error.
Heading error.
Roll (bank) error.
Pitch error.
Glide slope error.

This unit is designed for heavy aircraft that normally use a three-axis-of-freedom gyro. This gyro is placed near the center of gravity of the airplane. It can be shown that all movement of an aircraft in pitch, roll, or yaw is about the center of gravity; hence, the gyro is capable of very accurate response.

Assume that the pilot has selected a VOR frequency on the navigation receiver and a given radial on the course indicator. The switches shown in Fig. 12-3 are relay operated and will thus automatically be in the VOR position. This automatic switching is accomplished by a relay actuating voltage when a VOR frequency has been selected.

The steering pointer signal is a summation of several signals originating with basic attitude, azimuth, and positional sensors. This signal is made up of the algebraic summation of roll command and roll angle signals. The steering pointer receives its power from the output amplifier shown. The pilot, in controlling the airplane to keep the steering pointer centered, is essentially following a roll error signal. The degree of deflection of the steering pointer is a measure of how far and in what direction the pilot should roll the aircraft to make good his course. This pointer is pivoted at the bottom of the approach horizon to command a rolling movement pictorially.

Roll command is limited by a "bank limiter" to a predetermined value. This limit will have two values—one for landing approach and one for VOR navigation.

The roll command signal is composed of navigation error signal and an inner rate or damping signal. The navigation error signal is amplitude limited in the VOR mode to prevent circling of the aircraft. This is called *course cut limiting*. Its value determines the maximum cut that the aircraft will make at the course. The damping signal is the heading plus heading error passed through a low-pass filter. This low-pass filter has a time constant equal to 200 seconds. This combination of damping signals is identical to passing the heading signal through a high-pass filter. If the radio deviation signal exceeds two dots, the off-course relay operates and opens the heading signal to the 200-second low-pass filter. This prevents the filter from storing bad information. System gains are so adjusted that two-dot operation of the relay allows the system to compensate automatically even for severe crosswinds. The system is designed so that switching occurs without steering pointer transients.

For localizer approaches, the pilot need only switch to a localizer frequency. The direct heading feed is then automatically replaced with a derived damping signal. This damping signal is the summation of the best frequency components from roll, heading, and radio signals that are passed through a complementary filter. The damping signal is completely independent of static attitude, heading, or position of the airplane, and thus provides automatic crosswind correction. To prevent circling for large deviation signals (initial entry into the course) the high-pass section of the complementary filter is omitted. The omission of the high-pass section increases system stability at large distances from the localizer transmitter yet with adequate stability to approximately 1.2 miles from the localizer transmitter, which is the normal touchdown point for long runways used by high-performance aircraft. Complete usage of the complementary filter provides adequate stability well past touchdown (D = 0.34 mile), as well as providing the automatic crosswind feature. This switching is done automatically without objectionable transients in the steering pointer operation.

Figs. 12-6 and 12-7 show an exploded view of the course indicator and flight director horizon, respectively. Each instrument is a carefully worked out arrangement of galvanometer movements, servo motors, synchros, etc.

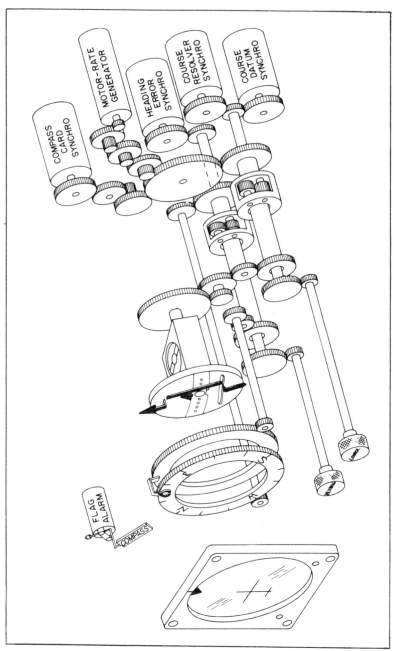

Fig. 12-6. Mechanical layout of Collins Model 331A-4 course indicator.
(*Courtesy Collins Div., Rockwell International*)

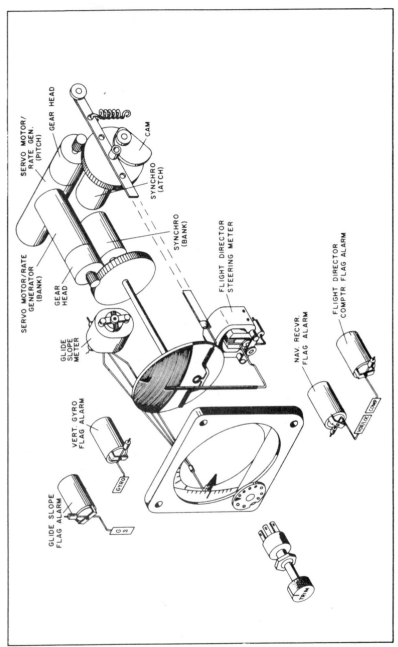

Fig. 12-7. Mechanical layout of Collins Model 329B-4 flight director horizon.
(*Courtesy Collins Div., Rockwell International*)

FLIGHT SIMULATORS

The first flight simulator was the Link Trainer that came into use before World War II. Since that time, advances in electronics and the computer art have made it possible to almost exactly simulate the control and operation of any airplane or space vehicle, although usually no attempt is made to simulate actual motion; that is, develop actual inertial forces. The simulator has two important advantages:

1. It is much less costly to operate than an actual airplane.
2. Practice conditions may be simulated that are too dangerous to set up in the air.

There are other advantages. A simulator may be operated at any convenient time of the day or night, it need not be near an airport,

Fig. 12-8. Desk-top Pacer MKII flight simulator. (*Courtesy Pacer Systems, Inc.*)

and it is more convenient to teach procedures on a simulator than when in the air. The student is alone in the cockpit, and the instructor communicates using an audio system. When a flight crew is instructed by simulator, the copilot, flight engineer, or others may actually operate as a flight deck team.

A flight simulator is designed around a set of differential equations that describe a specific aircraft. The equation for the roll axis was previously discussed. There will be a similar equation for the pitch and yaw axes, and the equations are interrelated. A computer is programmed for these equations. This process provides the flight attitude for the simulated airplane, which is displayed on cockpit instruments exactly as they are in an actual flight.

An internal clock relates the flight to distance, and direction is simulated by a linkage to the yaw or roll equations. The dynamic location of the simulated airplane on the X, Y, and Z axes can be used to simulate altitude, DME, VOR, and other navigational readings.

Compact flight simulators are now available for desk-top use. The Pacer MKII, shown in Fig. 12-8, is capable of simulating a full IFR flight of an airplane equipped with dual VOR, DME, ADF, transponder, and dual communication. The simulated airplane is a single engine type of less than 200 horsepower with flaps and retractable landing gear. A simpler version may be made up by ordering less options. The system is designed to begin with the Basic Flight Unit, which can be used to teach instrument maneuvers such as climbing and level turns, patterns, etc. Options are set up as follows:

Basic Flight Unit
    Primary instruments (including altimeter with Kollsman Window).
    Audio system.
    Transponder.
    Fuel.
    Takeoff and landing mode and engine system.
    Clock.
    Ignition assembly.
    Elapsed time meter (Hobbs).
    VHF COM.
    Environmental controls.
    Trim assembly.
Supplementary Flight Module
    Gear assembly.
    Flaps assembly.
    Gear and flaps position indicator.
    Rudder assembly.
ADF Module
    ADF digital frequency control.
    ADF bearing indicator.
    ADF audio mixer.
ILS Module
    ILS NAV control.
    ILS indicator (CDI and glide slope).
    VHF/ILS mixer.
    Marker beacon.

VOR/DME Module
    VOR/DME NAV control.
    VOR indicator (CDI and OBS).
    VHF/VOR audio mixer.
    DME indicator.

Systems such as this have wide application in secondary school programs and colleges as well as flight training centers. These devices impart the feel of the controls to the beginner, yet provide challenging situations for advanced training.

To be used effectively, training programs now called *software* or *courseware* have been developed that formalize simulator time and are more efficient to use. On the other hand, there is serious consideration to the use of simulators as "coin-operated" machines for the casual user. In this case, the machines will actually be turned on for a specified time by the insertion of dollar bills, rather than coins. The user receives private training, but does not receive "logged" time under a formalized program.

13

Autopilots

Autopilots are precisely what their name implies. They are capable of automatically holding the aircraft on course without human intercession. An autopilot can usually execute certain routine maneuvers better than a human pilot. The purpose of an autopilot is to relieve the pilot of controlling the airplane while engaged in such duties as navigation, communication, or record keeping. The autopilot also reduces pilot fatigue from long periods of flight "on the gauges," when the lengthy tedium of maintaining attitude by instrument reference is tiring.

The precise flight of an aircraft under autopilot control is from the ability of gyros to perfectly sense attitude change. Reaction to attitude changes depends on the characteristics of the airplane and speed. In the more sophisticated autopilots, a computer converts gyro sensing into control movements. The computer solves differential equations that satisfy the characteristics of a certain aircraft. Hence, the FAA requires that an autopilot be certified for each type of aircraft in which it is to be installed. More advanced autopilot systems are more properly called *guidance systems*. They do more than maintain the aircraft on a preset course. These systems automatically intercept a preset omni bearing and turn the aircraft into the heading, correct for crosswind, hold the aircraft on a localizer during landing, and provide other functions.

SIMPLE STABILITY SYSTEMS

It has been pointed out several times in this text that without gyros an airplane cannot be flown in conditions that obscure an attitude reference for the pilot. Skilled instrument pilots interpret gyro indications and correct the controls for normal flights. Unskilled pilots may rely on a stability system (a simple autopilot) to prevent deadly spirals when bad visibility is accidentally encountered.

The Brittain Model CSA-1 Stability Augmentation System is certified for several types of light aircraft. This system is unique in that a single gyro is used to sense both yaw and roll. It is done by mount-

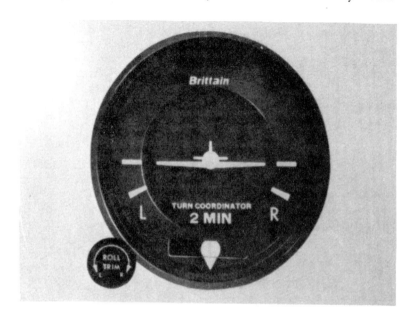

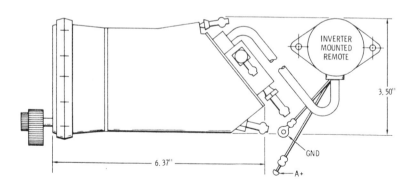

Fig. 13-1. Panel and side views of the Brittain Model TC-100 gyroscopic turn coordinator used with the Brittain Model CSA-1 Stability Augmentation System. (*Courtesy Brittain Industries, Inc.*)

ing the gyro in a direction that is not parallel to either axis. The panel-mounted, off-center gyro is shown in Fig. 13-1. Basically a rate gyro, the instrument senses rate of both roll and turn. The airplane will be in stable flight when both of these rates are zero.

The gyro may be driven electrically or by vacuum and replaces the rate-of-turn gyro that has been standard for many years. For electrical operation, a small inverter produces the 400 Hz required for the rotor. A patented pneumatic metering-valve system varies the vacuum applied to the servo actuators. Gyro displacement moves a spool-sleeve rotary valve. Rotation of the spool within the sleeve is proportional to roll or yaw rate. Air movement is to a point between vacuum supply and servo input. Vacuum differential positions the actuators. The servo actuators are shown connected to aircraft control wires in Fig. 13-2. These actuators are simple suction pistons that apply just enough force to move the aileron controls of a light aircraft. In the event of malfunction, the pilot may easily overpower these actuators. The ROLL TRIM knob provides command trim function. Clockwise rotation initiates right trim and vice versa. A cutout valve on the instrument panel deactivates the system by removing vacuum to the actuators.

This system is one of the least expensive autopilots obtainable and greatly increases the safety and convenience of an aircraft. It is easily

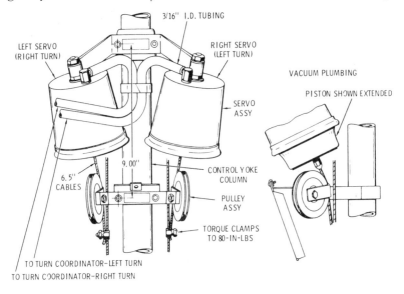

Fig. 13-2. Servo actuators of the Brittain Model CSA-1 Stability Augmentation System. (*Courtesy Brittain Industries, Inc.*)

installed by a certified mechanic from simple instructions. The usual precaution is that no accidental obstructions may arise from any condition that could jam primary controls. Neoprene tubing carrying vacuum (plumbing) must be free of kinks and away from heat.

THREE-AXIS AUTOPILOTS

The transition from flight director to autopilot is simply a matter of removing the pilot from the control loop. The pilot is replaced by servo motors (Fig. 13-3) that are attached to the control cables.

(A) *Servo motor drives control cables.*

(B) *Servo amplifier with cooling fins.*

Fig. 13-3. Servo motor and servo amplifier. (*Courtesy Astronautics Corp. of America*)

The first step in automatic control is stabilization of the roll/yaw axes. The next step is to control the pitch axis. The final step is to "couple" the autopilot to navigational and altitude sensors.

As discussed previously, gyros are capable of deriving two useful stabilization parameters. A gyro mounted in a three-axis gimbal will sense the instantaneous attitude. A rate gyro will sense the rate of change in attitude. A gimballed gyro can also be used to sense rate of attitude change if a means is provided to differentiate changes in attitude. A good method, however, would be to combine an attitude gyro to sense attitude and a rate gyro for use in computing the asymptotic return to true course. To put it another way, the rate gyro may be used for stabilization on a flight path determined by the attitude gyro.

Fig. 13-4. The control head of the Pathfinder autopilot. (*Courtesy Astronautics Corp. of America*)

The Pathfinder autopilot (Fig. 13-4) is an example of a system developed on a modular basis, using both attitude and rate gyro sensing. Modules may be added to provide additional degrees of capability. Beginning with rudimentary roll/yaw stabilization, the Pathfinder autopilot may be expanded to full capability, such as altitude hold and glide slope approach. The system expands according to the following breakdown:

Roll/Yaw Stabilization

A tilt-axis rate gyro (combined with turn coordinator display) is mounted on the instrument panel (Fig. 13-5). It measures rate of roll and yaw and provides the basic roll/yaw stabilization in all systems. It provides roll trim and standard rate turn sensing.

Heading Control

In P-1, P-2, and P-3 systems, roll/yaw excursions of the aircraft are integrated and are combined with the basic stabilization signal to maintain a constant heading. Heading changes are made with the TURN command knob of the turn coordinator.

Heading Select and Control

In P-2A, P-3A, and P-3B systems, heading signals from a directional gyro (slaved or unslaved) (Fig. 13-6) or a horizontal situation indicator are combined with the basic stabilization signal to provide heading select and heading hold.

Navigation Control

P-2 and P-3 systems have provision for VOR and localizer error signals to be combined with the basic roll/yaw stabilization signal to provide VOR/LOC tracking with crosswind compensation. P2A and P3A systems have the additional capability of providing automatic intercept, automatic capture, and track of any VOR radial or localizer from any heading.

Fig. 13-5. Turn coordinator. (*Courtesy Astronautics Corp. of America*)

Pitch Stabilization

A rate gyro in the pitch/altitude sensor stabilizes the aircraft about its pitch axis at whatever attitude the aircraft is trimmed for. Therefore, the aircraft's own trim system (electric or manual) becomes the pitch command control in the P3, P3A, and P3B systems.

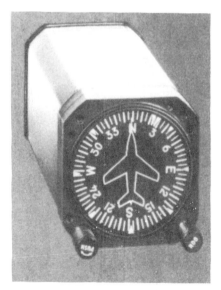

Fig. 13-6. The directional gyro of the Pathfinder is contained in this instrument. (*Courtesy Astronautics Corp. of America*)

Altitude Hold

An aneroid bellows signals changes from the altitude at which the "altitude hold" function is engaged. These signals are combined with the basic pitch stabilization signals to hold altitude. Any required trim changes are indicated on the trim meter and may be corrected manually (or automatically, if the electric pitch trim system is installed).

Glide Slope

The P3B system includes electric pitch trim and a glide slope coupler as standard. Automatic capture of the glide slope is provided by automatically substituting the glide slope error signal for the altitude signal when the glide slope is intercepted. At the same time a pitch-down signal is introduced to smooth glide slope capture. Glide slope can also be manually engaged.

The complete Pathfinder system divides into roll/yaw axes control and pitch axis control. Sensing elements for roll/yaw are:

Rate-of-turn gyro.
Directional gyro.
Radio (VOR and localizer) input.

Pitch axis sensors are:

Rate-of-pitch gyro.
Pitch platform.
Altitude hold bellows.
Radio (glide slope) input.

The sensors of the Pathfinder system use an illumination source for pickoff. This eliminates any mechanical coupling, hence, friction or stickiness between the gyro mounting and aircraft platform. Fig. 13-7 shows how a light mask changes position as the relationship between gyro and aircraft platform is displaced. The light source is interrupted as a function of gyro sensing, and the resultant light is picked up by a photocell, yielding an analog current. The same principle applies to the altitude hold bellows shown in Fig. 13-8. The bellows senses air pressure change due to altitude change and likewise displaces a light mask.

Up/down and right/left sense is converted to a dc analog as shown in Fig. 13-9. The simplified schematic is the yaw/roll function. The rate-of-turn gyro functions to displace the light mask and cause a photodiode to act whenever the aircraft yaws. When a right turn is commanded by rotating the turn knob, the switch is closed on a negative polarity, and CR1 conducts through R1 and R3. This acts as a voltage divider, applying a voltage to the noninverting input of

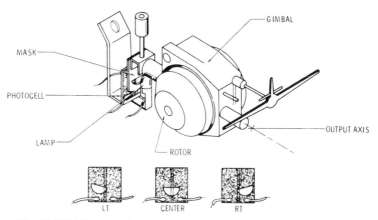

Fig. 13-7. Light mask changes position as the relationship between gyro and aircraft platform is displaced.

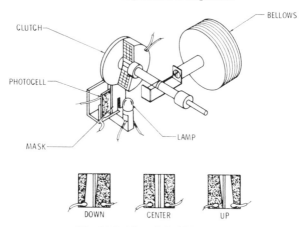

Fig. 13-8. Altitude hold bellows.

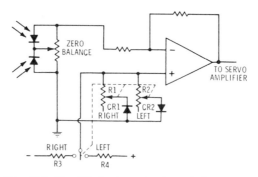

Fig. 13-9. Simplified schematic of yaw/roll function.

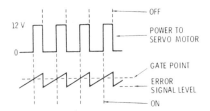

Fig. 13-10. Relationship of error signals to the final current furnished the motor.

the op amp. This places the aircraft in a roll/yaw attitude and displaces the rate-of-turn gyro. The system then zeros out when the right photodiode output voltage is equal and opposite to the amount set in. At this point the aircraft will be in a steady turn at a rate depending upon the setting of R1. For a left turn, the action is converse.

The Pathfinder uses dc motors to servo-actuate the controls. Torque furnished by the motor (hence, degree of control movement) is determined by a system of pulse-width modulation. Fig. 13-10 shows the relationship of error signals to the final current furnished the motor.

Duration of the pulse determines the amount of driving power. In other words, a wider pulse contains more energy. As the error signal level is moved up and down, the sawtooth intercepts the gate point at different times, thus changing the pulse duration.

The servo motors operate at just over 100 watts, hence, they are not powerful, but they furnish sufficient torque to move the controls of a light aircraft under normal conditions. The control cables are actuated from capstans. In an emergency, a pilot may easily override the autopilot servo.

AUTOPILOTS FOR HIGH-PERFORMANCE AIRCRAFT

Airliners and other high-performance aircraft require an autopilot that will react very rapidly with allowance for specific aerodynamic factors. The control head for a Collins AP-103 autopilot is shown in Fig. 13-11. In the following discussion of the AP-103, refer also to the block diagrams shown in Figs. 13-12, 13-13, and 13-14.

When the power switch is first moved to the ON position, all channels remain disengaged, but the pitch command is continuously synchronized with the pitch attitude of the aircraft. The pitch-control system is initially engaged along with the roll and yaw channels by depressing the ENGAGE button. Electrical interlocks prevent engagement until the vertical reference gyro is up to speed and system requirements have been satisfied. Initial engagement is possible only in the GYRO and HEADING modes. After engagement, the pitch-command synchronizer is disabled and the system accepts pitch commands from the pilot's pitch knob.

297

The altitude mode is engaged by moving the ALTITUDE switch to the ON position. Electrical interlocks prevent engagement of altitude-hold function unless the system is engaged. When the pilot's pitch-command knob is moved or when the approach (APPR) mode is engaged, the ALTITUDE switch returns to the OFF position. After altitude engagement, the system accepts ac error signals from the baro-metric altitude sensor.

Fig. 13-11. Control head of the Collins AP-103 autopilot. (*Courtesy Collins Div., Rockwell International*)

When the aircraft is at holding altitude, the pitch-command syn-chronizer circuit operates to eliminate steady-state altitude errors caused by trim changes. The time constant of the synchronizer in this mode is adjusted so that the basic altitude response is not affected.

The approach mode is engaged automatically if the automatic glide-slope capture circuit is armed. Arming is accomplished during the localizer phase of an instrument low approach by moving the function selector switch from the NAV/LOC to the APPR position. The ALTITUDE switch can be ON or OFF as desired. Actual approach en-gagement takes place when a preset glide-slope deviation is reached. This trip point is adjustable from 5 dots below to 3 dots above the glide path as revealed on the instrument panel. This depends on the approach characteristics of the aircraft and the desired bracketing maneuver. When the approach mode is engaged, the pitch-control system accepts signals from the glide-slope receiver. The ALTITUDE switch, if ON, is returned to OFF. In the approach mode, the pitch-command synchronizer again functions to eliminate steady-state glide-slope errors caused by trim changes. The time constant of the synchronizer in this mode is adjusted so that the basic glide-slope

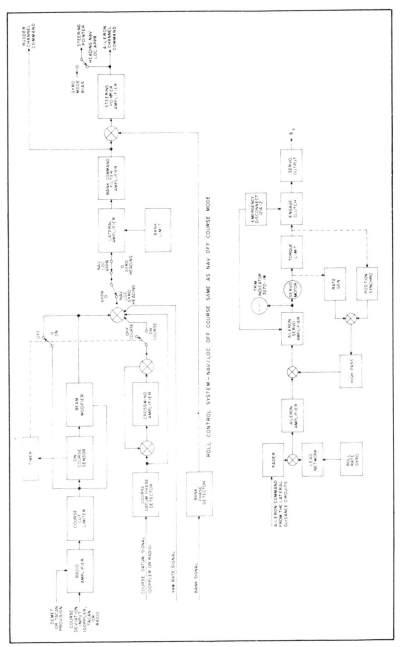

Fig. 13-12. Block diagram of the Collins AP-103 autopilot roll-control system, navigation mode. (*Courtesy Collins Div., Rockwell International*)

response is not affected. A preset switch bias is introduced at the time of automatic approach engagement.

The basic amplifier in the pitch-control system is a Type W magnetic dc amplifier that has provisions for mixing two isolated dc input signals and provides an isolated dc output. All ac signals are fed to phase detectors for conversion to dc prior to amplification.

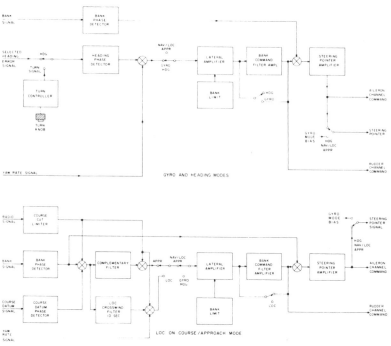

Fig. 13-13. Block diagram of the Collins AP-103 roll-control system, gyro heading localizer and approach modes. (*Courtesy Collins Div., Rockwell International*)

Two amplifier stages are used ahead of the servo power amplifier. One amplifier stage mixes and amplifies the pitch and up-elevator signals, a second amplifies either glide slope or altitude, and a third mixes the outputs of the two preceding amplifiers with the pitch-rate signal and amplifies the difference. Each amplifier has available an additional feedback winding for controlling its gain, thus providing a versatile system for signal shaping and gain scheduling as required for a particular aircraft.

When the dc pitch-range error signal reaches the power amplifier, it is mixed with a position and rate feedback signal from the servo. The servo position signal is passed through a high-pass network be-

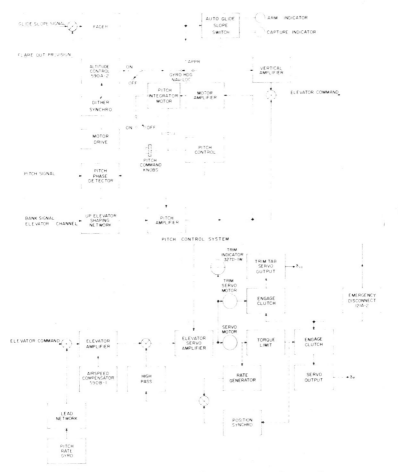

Fig. 13-14. Block diagram of the Collins AP-103 pitch-control system.
(*Courtesy Collins Div., Rockwell International*)

fore it reaches the mixing point to remove the low-frequency components. A high-passed component of pitch rate is added to pitch rate
to complement the lows removed from the position signal. These two
signals can be made to complement each other exactly in a theoretical system to give exactly the same result as a straight servo position
plus a pitch-rate system that is perfectly coupled to the control surface. The disadvantage of a straight position servo system is that the
capstan position becomes a poor approximation to pitch acceleration
at low-frequency, small-amplitude excursions. This is because surface friction and control-cable stretch tend to decouple the servo
from the surface under these conditions. This is particularly true

when the variations of surface effectiveness over the range of flight conditions is considered.

The complementary technique just described overcomes this disadvantage by removing these undesirable feedback components. Thus, the servo tends to produce not merely capstan deflections, but commanded pitching moments on the airframe, thereby producing optimum aircraft control. The degree to which this optimization can be realized depends on the amount of acceptable stick activity.

The signal output from the servo amplifier is fed directly to the trim indicator and to the trim-tab servo for automatically trimming the aircraft in pitch.

All the controls necessary to activate the roll-control system are contained on the autopilot controller. Since the roll and pitch channels are engaged simultaneously, the conditions for initial engagement are as described in the previous section. The function of each control is described for the various modes of lateral guidance in the following sections.

Gyro Mode

In this mode of operation, the lateral control can be considered as "power steering plus stabilized heading." Prior to engagement, a heading synchronizer drives a control transformer to follow the compass heading of the aircraft. After engagement, the synchronizer drive is disabled and the system accepts the heading-error signals from the control transformer, thus maintaining the heading existing at the time of engagement. If the aircraft was executing a turn at engagement, it will roll out smoothly and reacquire this heading.

When the pilot desires to maneuver the aircraft laterally, he uses the TURN knob on the autopilot controller to command a bank angle. As the TURN knob leaves the detent position, it activates the heading synchronizer again, the system accepts signals from the TURN knob, and as the aircraft rolls smoothly into a turn, the synchronizer follows heading changes. When the desired heading is reached, the pilot centers the TURN knob and a new stabilized heading is established.

Heading Mode

When the pilot desires to fly a selected magnetic heading, he moves the function selector switch from GYRO to HEADING. The fader circuit smoothly fades out any existing command signal, then switches the system to accept signals from the selected heading input. The TURN knob and heading synchronizer are disabled. If the pilot has preselected a new heading, the new heading command is faded slowly into the aileron channel and the aircraft responds by rolling smoothly into a turn toward the new heading, and rolling out smoothly when the new heading is acquired. Coordination is maintained in the other channels during this maneuver. Since all commands go through the smoothing filter, the pilot can make large abrupt changes with the heading selector without causing undesirable roll accelerations and rates.

NAV/LOC Mode

Two types of guidance are possible in the NAV/LOC position: (a) far-out localizer tracks, and (b) VOR radials.

LOC Mode—As the desired localizer beam is approached, the pilot sets up the localizer course, then switches the function switch to NAV/LOC. Fading is as described previously. The aircraft is then flown automatically to the center of the beam with the proper correction for crosswind automatically set in. The crosswind filter is complemented with radio to improve stability. The localizer mode consists of "off-course" and "on-course" phases.

In the off-course phase, the crosswind correction circuit is inoperative and the course-cut limiter determines the flight path toward the course. As the aircraft approaches the center of the beam, the on-course sensor switches to the on-course phase, activating the crosswind filter.

VOR Mode—The VOR mode has two phases, off-course and on-course. The off-course phase is similar to the localizer off-course phase except that the input radio signal becomes proportional to aircraft deviation from the selected VOR radial.

When the on-course sensor switches to the on-course phase, the beam noise suppressor (beam modifier), cone sensor, and crosswind corrector become operative. Time constants are switched to values suitable for VOR operation. With excessive beam noise or during flight through the VOR "cone of confusion," the radio signal is removed and the autopilot continues to fly its present heading with crosswind memory.

For "over-the-cone" operation, a logic circuit is activated to approximately determine the length of flight through the cone. Upon leaving the area of the cone, the logic circuit initially activates the off-course phase to permit course capture in the case where the pilot has selected a different outbound course. The logic circuitry is limited so as to cause the system to revert to the off-course phase anytime that the radio signal has been cut off by the beam noise sensor for a predetermined length of time. (A pilot may change course prior to entry into the cone, which will result in radio cutout.)

Approach Mode

The approach mode is engaged automatically, as previously described. In the roll-control system, the command smoothing filter is removed and a complementary damping filter is added. This complementary filter uses the best components of the radio signal, bank signal, and heading-error signal to derive a damping signal that is free of radio noise and long-term errors caused by crosswind.

All Modes

The command smoothing filter is common to all modes except approach. The general circuitry is similar to the pitch-control channel. The magnetic amplifiers are identical and interchangeable with those in the pitch-control channel. The aileron servo system is identical to the elevator servo system.

In addition to the aileron-command signal, there is a signal provided to the yaw-control system to establish coordinated turns.

Yaw-Axis Control

The yaw-axis control is energized upon engagement of the autopilot. The fader circuit accepts signals from the yaw-rate gyro and the rudder-washout integrator along with bank-command signals, and fades them smoothly into the rudder amplifier. The signal is then mixed with position and rate feedback signals from the servo and applied to the rudder servo amplifier to develop a control signal for the rudder servo. The output of the servo amplifier is also applied to the rudder-trim indicator.

Prior to autopilot engagement, servo-loop synchronization is provided to assure that the position synchro is at null and agrees with the streamlined rudder-control surface position (including proper trim). After autopilot engagement, this provides for servo system operation over the linear range of the position synchro. When mismatch of position synchro and control-surface position occurs, the rudder-washout integrator circuit automatically compensates the system. The system also provides the pilot with the capability for manually retrimming the aircraft rudder system without disengaging the autopilot. The rudder-washout integrator circuit integrates the rudder servo-amplifier output and produces a signal at the amplifier input to cancel both steady-state commands and rudder-position signals. The time constant of this integrator is of sufficient magnitude to maintain short-term yaw stability in the rudder channel.

14

Design and Reliability of Solid-State Avionic Systems

The advent of modern solid-state avionic systems has radically changed our approach to understanding and maintaining aircraft systems. Integrated circuitry no longer makes it possible to gain access to all points of a circuit with an oscilloscope probe. Nevertheless, although circuits have become more complex, they have also become more reliable than when thermionic emission was the only means of control. Vacuum tubes had a finite, often erratic life; tube equipment was bulky and generated unwanted heat. Semiconductors have now added a great deal to all characteristics of aircraft systems.

The great body of knowledge gained in vacuum-tube electronics has been applied to semiconductor electronics, and a new technology with new terminology has evolved. In many cases, with integrated circuitry, it is neither possible nor necessary to examine the circuit itself. Rather, the device is treated as a "black box" with known input and output parameters. The important thing is to deal with the structure of a solid-state system based upon environment and parameters. In this chapter we will examine some facets of solid-state electronics from the standpoint of reliability, physical construction, and electrical behavior as related to airborne electronic systems.

SOLID-STATE DEVICES

The term *solid state* is employed to signify that electronic action, that is, the theoretical flow of electrons, takes place in solid matter rather than gas or vacuum. Actually, matter itself is theoretically comprised of vast spaces in which mass particles circulate.

A modern electronic system is physically made of two classes of solid-state devices—the discrete device and the integrated circuit. A discrete device is a separate resistor, transistor, diode, and so on. An integrated circuit (IC) is a small sealed unit, sometimes called a *chip*, containing a number of devices arranged in a circuit to perform a specialized function. A device may be *passive* or *active*. An active device adds power to the circuit in a certain manner or otherwise performs some control function. A transistor is an active device. It receives power and adds it to the circuit. Resistors, capacitors, and diodes are passive devices. They do not add power to the circuit.

Resistors and capacitors are mass-produced by highly specialized machinery. The production process is such that the devices thus produced are tested and separated into values with specified tolerances, usually 5% or 10%. Sizes and ratings are standardized and values marked. Resistors are marked with the well-known color code. Several means are used to mark capacitors.

A bonus result of the development of the transistor is the much lower voltage and power required to operate a transistor circuit. This has meant less failures of resistors and capacitors that are included in a circuit. The circuit designer, of course, chooses ratings that are well within safe limits.

Semiconductor devices are also produced by highly specialized machinery, but the production process is much more complex, especially in the case of integrated circuitry. Great lengths are employed to ensure that no contamination enters the process. Here, too, however, transistors are tested and separated according to characteristics. A production process will have a certain "yield" for each class of parameter. A selection process is employed, in which out-of-tolerance units are reclassified. Devices employed for avionic circuitry usually meet the tightest specifications for their type of circuit employment.

THE SEMICONDUCTOR JUNCTION

Active solid-state elements, as well as diodes, pass current in a certain manner, and the applied voltage does not linearly relate to the current. These are semiconductors; they do not conduct in accordance with Ohm's law. That is, the current they pass is more a function of the current itself. The semiconductor is neither a true conductor as is a metal, nor is it an insulator; it lies somewhere in between, depending upon its composition. Germanium and silicon form the basis of most semiconductors now in use. A glance at Table 14-1 shows the difference in conductive properties between conductors, semiconductors, and insulators. The table shows values for pure crystalline germanium (chemical symbol Ge) and pure silicon (chemical symbol Si). In order to cause these elements to behave properly for use in electronics, certain "impurities" are carefully added in very minute amounts. These precisely added impurities decrease the resistance by making free electrons available.

Table 14-1. Resistance of Conductors, Semiconductors, and Insulators

Material	Resistance In Ohms per Cubic Centimeter	Category
Silver	10^{-6}	Conductor
Aluminum	10^{-5}	
Pure Germanium	50–60	Semiconductor
Pure Silicon	50,000–60,000	
Mica	$10^{12}-10^{13}$	Insulator
Polyethylene	$10^{15}-10^{16}$	

Semiconductor action is explained by the concept of "holes." That is, certain atoms making up the conductive materials lack an electron, thus creating a hole. These holes move about and can be considered as actual entities. Action within the semiconductor is then considered as the drift and flow of electrons and holes. The electron is classically considered as an electrical charge of precisely defined negative magnitude and mass. The hole is considered an equal opposite or positive charge, but, of course, does not possess mass. Once a hole accepts an electron, that atom becomes neutral in the conductive process. Certain impurities make free electrons available and others make holes available. Atoms with holes are known as *acceptor* atoms and those providing free electrons are known as *donors* (Table 14-2). A semiconductor which contains acceptor atoms is known as *p-type*, signifying positive charges or holes. Conversely, an *n-type* semiconductor contains donor atoms or negative charges. During formation of the semiconductor crystal, it can be made either p-type or n-type by careful addition of specific impurities that determine semiconductor type.

The purpose of this discussion is to acquaint you with factors relating to the action of semiconductors in terms of reliability and function in aircraft systems. This brings us to the basis of all solid-state circuitry, which is the semiconductor junction. A junction is formed when a p-type region and an n-type region are formed within the crystal structure. A terminal at the n region becomes the *cathode* and a terminal at the p region is the *anode*. Together, the device known as a *diode* is formed. It is the simplest form of semiconductor device.

Table 14-2. Function of Semiconductor Impurities

Semiconductor	Element (Symbol)	Function In Semiconductor
p-type	Boron (B) Aluminum (Al) Gallium (Ga) Indium (In)	**Acceptor** Each atom substitutes for a Ge or Si atom in the semiconductor crystal and can accept an extra electron, thus producing a hole.
n-type	Phosphorus (P) Arsenic (As) Antimony (Sb)	**Donor** Each atom substitutes for a Ge or Si atom in the semiconductor crystal and can donate an extra electron to the crystal.

When germanium is employed, current flows when junction potential reaches 0.2 volt. In a silicon device, current flow begins at 0.6 volt. These two values should be kept in mind. The junction tends to remain near this potential difference until current flow becomes too high, at which time the junction is destroyed by heat. The term *reverse current* describes current that flows when a relatively large opposite potential is applied. The point at which reverse current flows is sharply defined, and is known as the *avalanche* or *zener* point.

Fig. 14-1 shows the manner in which a semiconductor junction behaves to current flow when various potentials are applied. We have plotted applied voltage versus current flow. Note the potential at which current flow begins. This point is known as *forward potential*. As previously noted, it is approximately 0.2 volt for a germanium junction and 0.6 volt for silicon. The forward voltage for germanium tends to vary over a relatively wide range with temperature variation, whereas a silicon junction is more stable. Except in special cases, silicon semiconductors are preferred in avionic equipment.

As we pass through the forward potential, and *forward current* begins to flow, we note that resistance to junction current decreases. Although minute currents still flow below the forward potential, it is generally safe to assume that current is inactive below this point.

Diodes are used to perform many functions in avionics. They are used as ac rectifiers, clampers, switchers, and so on, and range from

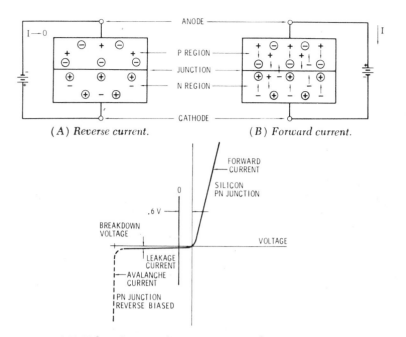

(A) Reverse current. (B) Forward current.

(C) Voltage/current characteristics of a silicon pn junction.

Fig. 14-1. The behavior of a semiconductor junction to current flow when various potentials are applied.

small signal carriers to large units carrying several amperes and requiring means of dissipating heat.

Silicon diodes can operate with upper temperature limits from 100°C to 200°C (212°F to 392°F). Temperature increase affects a semiconductor junction in two ways. First, the forward potential goes down (the figures, 0.2 volt for germanium and 0.6 volt for silicon, are averages), together with a decrease in forward resistance. Second, the minute reverse currents tend to increase (reverse current is often called *leakage current*).

THE TRANSISTOR

The transistor was invented by combining three semiconductors—one was a midjunction. Fig. 14-2 shows that the midjunction now becomes the *base* of a transistor. A transistor also consists of *emitter* and *collector*. The transistor shown is of the npn type, indicating that the collector potential must be positive and the emitter negative. When the transistor is a pnp type, the collector potential must be negative with respect to emitter and base. No current flows in the emitter-collector circuit unitil the base potential becomes approxi-

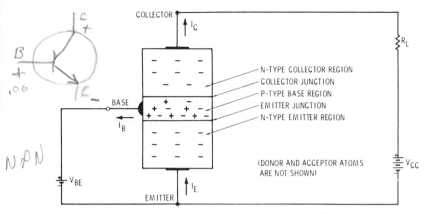

Fig. 14-2. Conduction in an npn junction transistor (common-emitter configuration).

mately 0.6 volt in the case of a silicon junction. The emitter-base circuit behaves as a diode. However, as current increases in this diode circuit, emitter-collector current also increases, but by a large factor known as *beta* (when employed in most circuits), which is usually greater than 10. This feature makes a transistor the control element of a circuit, acting as a switch or amplifier.

Circuit Configuration		Characteristics
Common Emitter (CE)	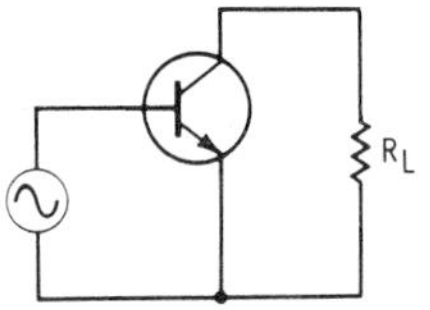	Moderate input impedance Moderate output impedance High current gain High voltage gain Highest power gain
Common Base (CB)	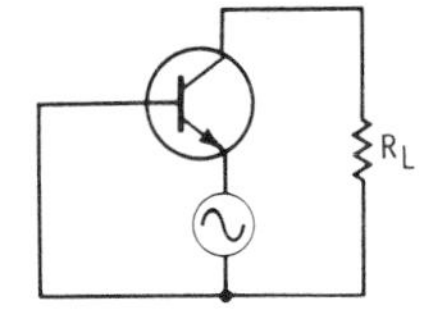	Lowest input impedance Highest output impedance Low current gain High voltage gain Moderate power gain
Common Collector (CC) (Emitter Follower)	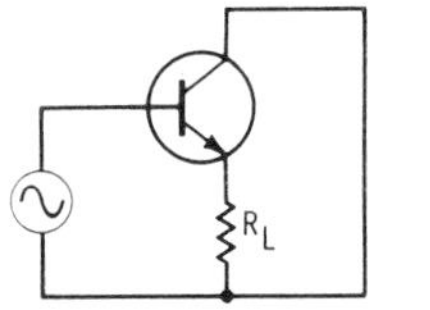	Highest input impedance Lowest output impedance High current gain Unity voltage gain Lowest power gain

Fig. 14-3. Transistor circuit configurations.

	Name	Symbol	Junction Structure	Range: Voltage-Current-Frequency	Application
Transistors	NPN Transistor	Base, Collector, Emitter	Base, Emitter, Collector, N P N	Small 0-1V / Large 1-10 Signal; Power to 100 W thru microwave frequency	Amplification Oscillation Power output Switching- detection
	PNP Transistor	Collector, Base, Emitter	Base, Emitter, Collector, P N P	Same as NPN	Same as NPN
	Unijunction Transistor (UJT)	Base 2, Emitter, Base 1	Emitter, P, Base 1, Base 2, N	Low power-under 1 watt; Low voltage-less than 100 V; LF 0-300 kHz	Trigger, Timing, Oscillator, Comparator — Circuits
	Field-Effect Transistor (FET)	Drain, Gate, Source	Oxide, Gate, Source, Drain, P	Power to 50 watts; Frequency to UHF; High Impedance; G_m to 20,000 μ mhos	Amplifiers-oscillators Switches-gates Constant I generator Voltage controlled R

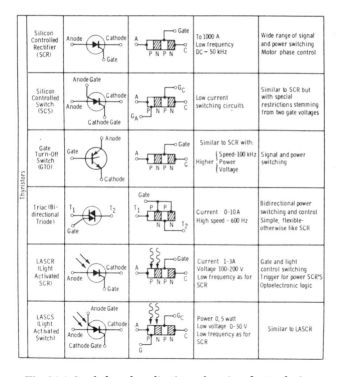

	Name	Symbol	Junction Structure	Range: Voltage-Current-Frequency	Application
Thyristors	Silicon Controlled Rectifier (SCR)	Anode, Cathode, Gate	A, Gate, P N P N, C	To 1000 A; Low frequency; DC – 50 kHz	Wide range of signal and power switching Motor phase control
	Silicon Controlled Switch (SCS)	Anode Gate, Cathode, Anode, Cathode Gate	A, G_C, P N P N, C, G_A	Low current switching circuits	Similar to SCR but with special restrictions stemming from two gate voltages
	Gate Turn-Off Switch (GTO)	Anode, Gate, Cathode	A, Gate, P N P N, C	Similar to SCR with: Higher { Speed-100 kHz, Power, Voltage }	Signal and power switching
	Triac (Bi-directional Triode)	T_1, T_2, Gate	Gate, T_1, P, P, N, N, T_2	Current 0-10A; High speed - 600 Hz	Bidirectional power switching and control Simple, flexible; otherwise like SCR
	LASCR (Light Activated SCR)	Cathode, Anode, Gate	A, Gate, P N P N	Current 1-3A; Voltage 100-200 V; Low frequency as for SCR	Gate and light control switching Trigger for power SCR'S Optoelectronic logic
	LASCS (Light Activated Switch)	Anode Gate, Cathode, Anode, Cathode Gate	A, G_C, P N P N, C, G	Power 0.5 watt; Low voltage 0-50 V; Low frequency as for SCR	Similar to LASCR

Fig. 14-4. Symbols and applications of semiconductor devices.

311

The transistor may be employed in three fundamental configurations as shown in Fig. 14-3. Notice the different characteristics for each configuration. This implies that the transistor is used as an amplifier. One of the first and most important uses of the transistor, however, was as a switch. This was as much due to its inherent limitations as to its strengths, since the transistor by itself is not capable of the subtle means of control that can be achieved by a vacuum tube. When used as a switch, the emitter-collector current is merely turned on or off by a base current sufficient to cause *saturation*. This is a condition wherein any increase in base current will not change the amount of emitter-collector current, which is at some maximum value as determined by the resistance in the collector circuit.

Once the transistor principle was perfected, many variations and additions were invented to perform a variety of control functions. Some of these are shown in symbolic form in Fig. 14-4. To become acquainted with the typical characteristics of the various semiconductor devices it is necessary to consult the manufacturer's specifications.

INTEGRATED CIRCUITS

The development of transistor fabrication processes led to the concept that semiconductors may be grouped together in many ways, forming what is in fact a whole circuit. Resistors can be created with a semiconductor layer, and junctions may be fabricated that perform the same function as a transistor or diode. By a number of methods, therefore, electronic devices may be created that perform in the same manner as an entire circuit made up of discrete devices. Because inductors and capacitors are very difficult to form by the processes of integrated circuitry, circuits are essentially made up by the lavish use of only diodes, transistors, and resistors. Capacitors of small values may be fabricated within the chip; however, when circuit elements other than transistors and resistors must be employed, they are usually placed outside of the chip by making leads available. In the manufacturing process, transistors and diodes are most easily formed. Resistors of defined value are somewhat more difficult to form. We can expect the circuit designer, therefore, to use elements with this priority. Schematics of integrated circuits appear complicated and difficult to interpret because of the profuse use of transistor junctions. It is usually unnecessary, however, to understand operation within an integrated circuit, since we deal with integrated circuits as single entities with input and output.

It has been previously mentioned that the most predictable behavior of transistors is as switching devices in which they operate in either cutoff or saturation. Switching circuitry is comprised of a family of ICs known as *digital* integrated circuits. These ICs operate by a combination of off/on or logic-0 and logic-1 conditions. A second family is classified as *linear* integrated circuits. Linear circuits are amplifiers made to perform in a number of ways. They are called *linear* because input and output magnitudes have a linear relationship.

The specific characteristics of integrated circuits can be obtained from manufacturers' specification sheets. Unfortunately, the specification sheet often implies familiarity with the terminology and highly specialized usage to which the given device is intended. It is therefore necessary to consult a text that covers the specific usage.

RELIABILITY OF AVIONIC SYSTEMS

The matter of reliability has undergone voluminous study in recent years, often with confusing results. One possible reason is that a system consists of elements of a wide range of behavior. A flight director system, for example, consists not only of the semiconductor devices that energize it, but also of the delicate mechanical components that comprise the instrumentation. Also to be considered are the connectors and interwiring.

When solid-state systems were first introduced, there was some regret that the reliability of the devices was less than hoped for. It now appears, however, that solid-state reliability has even surpassed expectations. Semiconductor junctions used in aviation systems now exhibit average failure rates of 0.01% per 1000 hours. Moreover, the devices do not "wear out" in the same way as vacuum tubes. In fact, semiconductors appear not to wear out at all. Failure in semiconductors (when it does occur) has been traced to physical and chemical processes associated with mounting and packaging. A great deal of development has gone into this aspect of manufacture.

The important independent variable in the determination of reliability is environment. The obvious environment of an aircraft system is in flight, taking under consideration vibration, temperature, air pressure, and so on. A less obvious environment is ground storage. Corporate and, to a greater degree, privately owned aircraft, spend a great deal of time in hangers or on tiedowns. In this case, temperatures within compartments may remain above 100°F for long periods. There is also the possibility of corrosive atmosphere.

Environment must be systematically analyzed from the standpoint of the following stresses:

Mechanical.
Temperature.

Electrical.
Other.

We may now proceed to further break down each of the foregoing. In each case, consideration must be given to both in-flight and ground conditions. Note that remarks apply to every component of a system considered by itself, even the smallest resistor.

Mechanical stress results from static force, shock, vibration, and air pressure. A static force is exerted by the means of mounting. It may lead to effects on those other categories of shock and vibration. These latter are interrelated. Mechanical shock results from inertia or pressure. Equipment is exposed to shock, twisting, and so on during installation and servicing. Exposure within the cockpit may result from in-flight mechanical shock from the flight crew. It is well to remember that even the mere removal of a connector is in fact a mechanical shock, even though made negligible through good design. Vibration is a series of shocks delivered at a certain frequency. Vibratory stress can only be treated by the designer and installer. Air-pressure stress obviously occurs with altitude changes. Semiconductor devices are hermetically sealed, but the seal may fail. Electrolytic capacitors may also be pressure sensitive. Pressure, therefore, is a stress that may only be considered by the designer or manufacturer.

The subject of mechanical stress, specifically shock, is important to the user, maintainer, and installer as well as designer. We speak of "wear and tear," which is, in fact, the many shocks that may occur routinely. Shock is taken here in its broad sense, meaning any momentary mechanical stress other than that brought about in the specific operation of a piece of equipment. It has been mentioned that the mere removal of a connector is, in fact, a shock. It is possible that the majority of failures result from shocks taken in this broad sense. For this reason, the troubleshooter should consider mechanical failure first. This would mean inspection of cables, joints, connectors, printed-circuit plug-ins, and so on.

The next environmental category is temperature, which has been briefly mentioned earlier. Temperature stress may be static, shock, or cycling. Static temperature stress results from long periods of sun exposure and during flight operation. High static temperature stress may result in electrical component failure. Plastic materials may have accelerated wear-out. The effect of contaminants is accelerated. Cold causes brittleness. Lubrication is affected. Temperature shock means sudden temperature change in either direction. Acceleration from hot to cold may result in mechanical breakage. Cycling from hot to cold and back results in thermal fatigue. Electrical components are most likely affected by thermal shock.

Electrical stresses are of four categories—voltage, current, continuous power, and cycled power. Voltage may exist with no current flow, but not vice versa. Voltage stresses may cause minute ionic (chemical) action such as corrosion. In semiconductor logic circuitry this could result in failure. More gross corrosion may also occur in connectors or near batteries. The effect of contaminants is accelerated by current flow. The effect of current stress is through the creation of magnetic lines of force, which, in turn, affect circuit action or induce corrosive voltages. Although both voltage and current stress have very illusive effects, *power* stresses are more obvious. Power produces heat which leads to temperature stress. Also, power applied to mechanical devices has obvious effects. The designer allows for this in rating the various system elements. Continuously applied power is the end usage of any system. It is the factor that leads to any failure due to normal use.

Cycled power occurs in start-up and stop, as well as transient conditions. Here is the reason for cockpit checklists that specify the order of turn-on or start-up. Transients are momentary surges in either direction produced by a variety of causes. The designer and installer must be aware of transient conditions that may lead to failure. When failure occurs, it will be in the component that is weakest at that moment.

Other stresses leading to failure are under such headings as humidity, corrosion, abrasion, and so on. The pilot, maintenance technician, installer, designer, and manufacturer all exert influence on these stresses. Abrasion, for example, may occur during manufacture as well as in operation. Again, we are considering each category and subcategory of stress as applying to every individual component or element of the system. This applies, for example, to the bearing of a servo shaft as well as to an obscure resistor in some logic circuitry. This, in turn, leads us to a discussion of actual failure analysis.

FAILURE ANALYSIS

Contributions to failure begin with the raw material itself and continue through final design, manufacture, and installation. Actual failure, however, occurs in the field. No device is absolutely failure proof. Rather, we approach the problem by attempting to arrive at the reasonable time that a failure may be expected. Reliability may be increased at greater expense, but the relationship is not linear. We reach a point where further expense is unwarranted.

Reliability may be defined as the *probability* that a device or system will operate under specified conditions within specified limits of accuracy or performance for a given time. Note that, strictly speaking, it is not necessary for the unit to become totally inoperative. In certified aviation electronic systems, the system is assumed to have failed when any part becomes inoperative or falls below performance specifications. There are many practical considerations to reliability.

A system may exhibit susceptibility to abusive manipulation that is apart from in-service use. For example, an autopilot system may require a certain type of periodic maintenance. This act of maintenance itself may, in turn, create stresses that lead to failures in a system that otherwise would not fail for a long time.

One way to measure reliability would be through actual operational testing. A problem here is that, for statistical accuracy, a great number of systems must be tested for a long time. Another method is to increase stress for a certain operational time on the theory that the increased stress will relate to in-service reliability.

Most systems will at least have been tested to some extent through actual operation, even though the tests are not statistically adequate. To be statistically adequate, testing costs on most aviation systems would be enormous. There is a condition known as *infant mortality*, whereby it is assumed that many failures occur during the initial operation of a new system. On this basis, all newly built systems are run for a period before installation, sometimes at increased stress. This is called *burn-in*.

The study of reliability is an actuarial study based upon probability theory. It is a mathematical treatment stating that all systems will fail some time, with the probability of failure increasing as the system operates longer. We have seen, however, that environmental stresses are present even when the system is inoperative. Nevertheless, operational time is the only reasonable parameter for mathematical treatment.

Reliability can be measured by the chances of failure at any point in the lifetime of a device or system. It is possible to chart this. The vertical axis will be probability of failure (one chance in six, one chance in three, etc.) and the horizontal axis will be the operating lifetime. It should be noted that workmanship failures occur first. This is why a manufacturer can honestly warrant a system. Beyond a certain time, wearout is the predominant cause of failure.

RELIABILITY VERSUS REPLACEMENT

Critical times in the life of a system are when the failure rate is increasing or decreasing. A logical time to replace would be just before failure rate increases, providing that this point can be determined. Ironically, we note that at certain points beyond this the failure rate decreases. This is the same as saying that having survived to time t_1, the chances are better of surviving to time t_2.

In the commercial airline fleet, utilization rates are very high. It is common for an airliner to spend eight hours of every day in flight. At this rate, hourly lifetimes and operating costs may be closely figured and component replacement rates assigned. When vacuum tubes were still in use, tubes were sometimes replaced and discarded at uniform periods.

The situation with general aircraft, and with solid-state systems, is more diversified. We have seen that nothing would be gained by the periodic replacement of solid-state circuitry. Unlike engines that may be periodically overhauled to reduce failure rate, it would be highly outrageous to overhaul most general aviation electronic systems as a means of failure prevention. The most likely wearout takes place in mechanical components associated with an electronic system, such as servo components, switches, and even potentiometers.

Excluding design deficiencies (i.e., poor component choice or functional concepts), mechanical shock, taken in its broad sense, probably contributes the most to electronic system failure. Since the shock is a random event not usually associated with actual operation, failure from this cause is difficult to predict. It can only be eliminated by eliminating the need for handling or manipulating equipment other than that of regular operation.

SOLID-STATE DISPLAYS AND READOUTS

The development of solid-state illuminated displays has resulted in the first significant change in cockpit instrumentation in several decades. The illuminated display has many advantages, particularly in displaying navigational information. Almost all of the communication and navigational units now produced use some form of illuminated digital display (Fig. 14-5).

Fig. 14-5. The Genave GA/2000 NAV/COM unit shown here uses a seven-segment tungsten readout that may be dimmed to accommodate variations in cockpit lighting. (*Courtesy General Aviation Electronics, Inc.*)

Most of the information to be displayed in the cockpit is numerical. A standard seven-segment method has been devised to form all numerals from 0 to 9, as shown in Fig. 14-6. Note that the numeral 8, as shown, uses the maximum number of segments. Any of the numerals may be formed by selecting certain segments for illumination.

The vertical segments are often slanted slightly. This rids the numeral sequence of boxiness and helps make characters stand out. Four methods have been devised to form the illuminated segments—light-emitting diode (LED), liquid crystal (LCD), tungsten filament, and gas discharge. In each case, the light source is designed as an oblong that forms the numeral segment.

The LED is a pn junction diode. Light is produced by the energy given up by the junction when current flows. When *gallium arsenide phosphide* (GaAsP) is used as the semiconductor, the emitted light is red or yellow. When *gallium arsenide* (GaAs) is used for the semiconductor, green light is emitted. The junction emits light when approximately 1.2 volts forward bias is present and about 20 milliamperes of current flows. This type of light source operates conveniently from a low-voltage source, and may be switched on and off rapidly with transistor logic.

There are two types of liquid-crystal displays, *dynamic scattering* and *field effect*. The dynamic-scattering type liquid-crystal cell is illustrated in Fig. 14-7. The liquid-crystal material may be one of several organic compounds in liquid form. This liquid is layered between glass sheets with transparent electrodes deposited on the inside faces. When a voltage is applied to the electrodes, charge carriers flowing through the liquid disrupt the molecular alignment and produce molecular turbulence. The liquid crystal is transparent when molecular turbulence is not present. When activated, the molecular turbulence causes light to be scattered in all directions, causing the cell to appear bright. This process is known as *dynamic scattering*.

A field-effect liquid-crystal display is similar to the dynamic-scattering type, but two thin polarizing optical filters are placed at the inside surface of each glass sheet. The liquid material in the field-effect cell is different from that employed in the dynamic-scattering cell. This liquid-crystal material "twists" the light passing through the cell when the cell is *not* energized, allowing light to pass through the optical filters, and the cell appears bright. When the cell is energized, no light twisting occurs and the cell remains dull. In another form, the cell can be made to appear bright when energized.

Liquid-crystal cells may be *transmittive* or *reflective*. In the transmittive type, both glass sheets are transparent, so that light from a rear source is scattered in the forward direction when the cell is

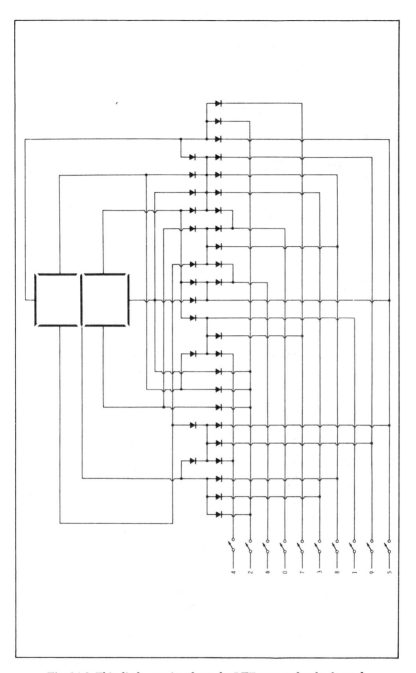

Fig. 14-6. This diode matrix selects the LED numeral to be formed.

319

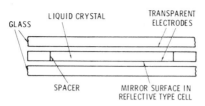

Fig. 14-7. Construction of a liquid-crystal cell.

activated. The reflective type has a mirrored surface on one of the glass sheets. In this case, incident light on the front surface of the cell is dynamically scattered by an activated cell.

Liquid-crystal cells cannot be energized by direct current. Current flowing constantly in one direction will cause plating of the cell electrodes with resultant damage. Hence, ac in some form is used. A great advantage of the liquid-crystal cell is the very small power required for operation. A small seven-segment dynamic-scattering liquid-crystal numeral requires only about 30 microamperes. A field-effect cell of the same size operates on as little as 300 microamperes. Despite the small current requirement, liquid-crystal displays offer bright light. The typical seven-element LCD produces bright light with only about 150 microwatts. Its chief disadvantage is that there is a comparatively long decay time after the element is switched off. At low temperatures this decay time is greatly increased. The eye can perceive the fadeout as segments are switched off.

Tungsten (or incandescent) numerical displays use small tungsten filament lamps placed behind lenses arranged in seven-segment format. Tungsten displays use more power and produce heat. An advantage is that they can be made very bright, yet they can be dimmed by reducing current.

15

Shop Facilities and Regulations

Standards for electronic systems have been established in many ways. National and international groups have been formed to link the world to a common system. Only by this means has it been possible to obtain the constant evolution of the industry. Persons engaged in aviation must be familiar with the various standards and the organizations behind them. The FAA regulates the certification of both repair shops and repairmen. Good maintenance and repair techniques are essential in aeronautical electronics, just as they are in any other branch of this field, but here details of testing are much more stringent. Shop standards and practices must be approved, and repair of aviation equipment is carried out under the highest standards of the industry.

STANDARDIZATION AND REGULATORY AGENCIES

Economical aeronautical operations require standardization of equipment, accessories, and methods. Safe aeronautical operations require regulations based on sound criteria. Standardization and regulation overlap in many areas. Throughout the years, a number of industrial, quasi-governmental, and governmental organizations have been formed to establish such standardization and regulation.

Radio Technical Commission for Aeronautics

Although organized in 1935 at the invitation of the United States government through the Department of Commerce, the Radio Technical Commission for Aeronautics (RTCA) is not an official agency of the government. Rather, it is a group of governmental agencies and industrial organizations, representatives of aviation and trade groups, and any others with vital interest in aeronautical electronics. Its purpose is to coordinate efforts in the technical development of electronic devices in aeronautics. Member agencies delegate such experts as engineers, pilots, and others to act on committees.

Usually, when a need exists or when some new development becomes feasible, the initiative is taken by one of the RTCA members. The group then sets up standards, after which it issues papers giving minimum performance standards, means of testing, and so on. After this is done, the FAA issues Technical Standards Orders (TSOs), which refer to the applicable RTCA paper as a standard. Thus the RTCA, in effect, sets the performance standards.

The RTCA also coordinates the views of government and industrial bodies concerned, and its recommendations are usually incorporated in the government regulations. Recommending frequency allocations and making studies on air traffic control, based on limitations of the electronic systems, are just two examples of the RTCA's many duties. Its headquarters is in Washington, DC. Governmental organizations that belong to it are the Departments of State, Treasury, Army, Navy, Air Force, and Commerce, the FCC, and the FAA.

Some of the private and industrial organizations that belong include Aeronautical Radio, Inc., Air Line Pilots Association, Air Transport Association of America, Inc., Aircraft Owners and Pilots Association, National Business Aircraft Association, and Electronic Industries Association.

International Civil Aviation Organization

The International Civil Aviation Organization (ICAO) is a specialized agency of the United Nations with headquarters in Montreal, Canada. It studies problems in international aviation, establishes international standards and regulations for civil aviation, and provides financial and technical aid for maintaining air navigation, airports, and ground facilities in areas where the populace is unable to maintain them. The ICAO is composed of 73 member nations and is headed by a secretary-general.

The ICAO maintains permanent technical committees that make recommendations. The committee on electronics, for example, performs functions which, on an international level, closely parallel those of the RTCA, and thus projects standards on an international level. The worldwide airways network utilizing VOR and similar navigational systems is the work of ICAO.

Aeronautical Radio, Inc.

Aeronautical Radio, Inc. (ARINC) is actually a corporation whose principal stockholders are airlines. Other stockholders are aircraft manufacturers, other transport companies, and foreign airlines. Headquarters are in Washington, DC, but ARINC also operates many worldwide ground stations that provide company communications channels for airlines. ARINC sponsors the Airline Electronic Engineering Committee (AEEC), composed of airlines engineering and technical personnel, and another committee for Europe, known as the European Airline Electronic Committee (EAEC). These two establish "Equipment Characteristics" which, in effect, are detailed standards that stipulate form factors, interconnections, tolerances, and other characteristics to show the electronic equipment manufacturer what the airlines are seeking and the standards desired. Although this does not mean that an airline will only purchase equipment built to meet ARINC standards, very little "non-ARINC" equipment is used aboard airliners.

Federal Communications Commission

The Federal Communications Commission (FCC) is an independent United States government agency that licenses all radio transmitting apparatus and determines the frequency bands and emission characteristics for each type of service. Because of the complexity of aviation communications, the FCC works closely with the FAA and RTCA. The aviation services are governed under Volume V, Part 87 of the FCC *Rules and Regulations*.

FCC REQUIREMENTS

In order to legally operate the transmitter of an aircraft radio, the pilot or any other person must possess a valid Restricted Radiotelephone Operator Permit, FCC Form 753-B, which any citizen can obtain from the FCC without taking an examination. This permit must be in the possession of all persons operating the equipment. Application forms are available at most airports. When purchasing an airplane, the new owner must apply for an Aircraft Radio Station License using FCC Form 404, even though the equipment was previously licensed. Applications for renewal of station license must be submitted on FCC Form 405-B. These forms can also be obtained at most airports.

An aircraft transmitter must be "type-accepted" by the FCC before the manufacturer can market it. Transmitters must have a frequency stability of $\pm0.005\%$. All transmitter adjustments or tests in connection with installation, servicing, or maintenance must be made by, or under the immediate supervision of, a person holding a First- or Second-Class Operator License—either Radiotelephone° or Radiotelegraph, depending on the type of equipment. The licensee is responsible for the proper functioning of the equipment, and for its compliance with all FCC requirements after such installation, repairs, or adjustments are made.

THE CERTIFIED REPAIR STATION

FAA rules governing the operation and certification of repair stations are set down in Civil Aeronautics Manual 52 (CAM 52). Any shop desiring to repair, install, or adjust aviation electronics equipment must apply to the FAA for certification with a radio rating that will allow the shop to diagnose malfunctions; maintain, repair, and alter equipment, including replacement of parts; inspect and test the equipment; make frequency checks and perform any calibrations necessary for proper operation of equipment.

A certification may have a rating of "unlimited," which means that a shop can work on all equipment of its class; or "limited," which will specify, by model or type number, the equipment on which the shop may perform repairs or adjustments.

To obtain a rating for a certain model or type of equipment, an applicant must show that he has the proper instruction manuals issued by the manufacturer, plus the prescribed test equipment. Electronic equipment is divided into three classes, each with job functions that have been standardized by CAM 52. An applicant must provide the personnel, equipment, and material required for competent performance. Job functions for the three classes are as follows:

Class 1. Communications Equipment

1. Diagnosis, as follows:
   a. Check wiring, antennas, connectors, relays, and other associated components to detect installation faults.
   b. Check engine ignition systems and aircraft accessories for sources of electrical interference.
   c. Check aircraft power supplies for proper functioning.
2. Maintain, repair, and alter radios, including installation and replacement of parts, as follows:

°Effective August 7, 1981, all First- and Second-Class Radiotelephone Operator Licenses are to be systematically phased out and replaced by a new class of license to be known as the *General Radiotelephone Operator License.*

a. Overhaul, test, and check dynamotors, inverters, and other rotary apparatus.
b. Paint and refinish equipment containers.°
c. Mark panels with calibrated settings and other information.°
d. Make and reproduce drawings or photographs to record alterations or modifications to radios.°
e. Make tuning-shaft assemblies, brackets, cable assemblies, and other components appropriate to aircraft-radio installations.°
f. Align rf and if circuits.
g. Test and repair headsets, speakers, and microphones.
h. Install and repair antennas.
i. Install complete radio systems in aircraft, and prepare weight and balance reports.°
(Qualified personnel must supervise alterations to the aircraft structure.)
j. Measure modulation, noise, and distortion.
k. Measure audio and radio frequencies.
l. Measure transmitter power output.
m. Measure inductance, capacitance, resistance, etc.
n. Measure parameters, determine characteristics, and provide proper locations for antennas.
o. Make operational tests with appropriate portable test apparatus.
3. Inspect and test radios, as follows:
a. Perform visual and mechanical inspection.
b. Perform electrical inspection with appropriate test equipment.
c. Test radio instruments.°
d. Test all types of tubes used in equipment appropriate to the rating.
e. Test resistors, capacitors, transformers, and other related items.
4. Make frequency checks by measuring radio frequencies to the required tolerances and calibrating the equipment to these tolerances where applicable.
5. Perform the calibrations that are necessary for proper operation of radios and apply to all preceding functions.

Class 2. Navigational Equipment

In addition to checking for faults in the installation:

° It is recognized that many operations can best be performed by specialized facilities. The job functions marked with an asterisk may be "farmed out" to specialized concerns.

1. Maintain, repair, and alter radios, including installation and replacement of parts, as follows:
   a. Measure loop-antenna sensitivity by appropriate methods.
   b. Determine and compensate for quadrantal error in ADF equipment.
   c. Measure rf attenuation in transmission lines.
2. Inspect and test radios, as follows:
   a. Calibrate ILS equipment to approved performance standards.
   b. Calibrate vhf navigational systems to approved performance standards.
   c. Calibrate vhf marker-beacon receiver systems to approved performance standards.
   d. Calibrate any navigational equipment, approach aids, or similar equipment appropriate to the rating.
   e. Determine waveforms and phase in radios when applicable.

Class 3. Radar Equipment

In addition to checking for faults in the installation:

1. Maintain, repair, and alter radars, including installation and replacement of parts, as follows:
   a. Provide equipment and materials to metal plate transmission lines, waveguides, and similar equipment, in accordance with appropriate specifications.
   b. Provide equipment and materials to pressurize appropriate radar equipment with dry air, nitrogen, or other specified gases.
2. Inspect and test radars by providing all equipment and materials, as listed for the same jobs under Class 1.
3. Perform the calibrations that are necessary for the proper operation of radars and listed for the same jobs under Class 1.

A certified shop must have proper temperature control and lighting, and adequate storage and work areas. It is the responsibility of the repair-station officials to employ competent personnel. Any individual in charge of inspection, maintenance, and overhaul must have practical experience of at least 18 months in all functions of the work he supervises. Although the management is responsible for the competence of supervisory personnel, the FAA may request the employer to submit resumés or work records on the background of supervisory or other key personnel. When applying for a repair-station certificate, the applicant must recommend one or more individuals at or above the level of shop foreman to act as certified repairman. Moreover, management must certify that these individuals are competent to perform and supervise the work to which they are assigned. The

FAA will then appoint these individuals as Certified Repairmen, in accordance with Civil Aeronautics Manual 24 (CAM 24) Subpart B, Par. 24.100, and only they may inspect and "sign off" finished work. Records must be maintained of the assignments of certified personnel and also of each job. The certificate of an individual is valid only during his employment, and only while assigned appropriate duties.

TECHNICAL MANUALS

A certified repair facility is expected to maintain a library of up-to-date technical manuals of the equipment it expects to repair or test. Manuals must be provided in order for certain equipment to be certified. Manufacturers provide such manuals either at a cost or as part of an agreement. Some equipment may require several manuals for complete coverage. In addition to the manuals intended for technicians, pilots' manuals may also be provided. These should also be available in shop libraries.

Literature for avionic equipment usually follows terminology and format adapted from military manuals. In general, a technical manual will be sectioned as follows:

- Introduction and descriptive material.
- Operating procedure.
- Theory of operation.
- Maintenance and troubleshooting.
- Parts list.

This arrangement is considered most logical, and suits the purpose for most equipment. The section covering operation usually duplicates the pilot's manual, although it may contain special instructions for bench operation. The "theory" of operation should be a functional description of the circuitry, sufficiently clear to provide knowledge for a technician to isolate trouble, repair, and test. The section on maintenance and troubleshooting usually also contains the step-by-step procedure for performance testing to ensure that equipment is performing according to specifications. In case of accident due to equipment failure, these procedures are among the only documentation to support proper operation or suspected fault. Specific models of test equipment called out in the test procedure are presumed unless the term, *or equivalent*, is added.

Factory training schools and factory representatives may also distribute valuable literature. In some cases equipment is introduced before manuals have been prepared. Changes to existing technical manuals are distributed, and the manuals must be kept up to date within the shop library. Such changes may affect flight safety.

The quality of technical manuals varies widely among the various manufacturers, even though the equipment itself is of high quality. Realistic troubleshooting procedures are difficult to prepare, particularly before production models have entered service. Manufacturers are sometimes reluctant to divulge functional details for proprietary reasons. Complaints of technical manuals range from skimpy, erroneous, hazy subject matter to poor printing and graphical presentation. The quality of literature is now superior to that of the past, however.

TEST EQUIPMENT

A large variety of excellent electronic test equipment is available, much of which can be readily used in maintaining and troubleshooting avionic equipment. For troubleshooting, any oscilloscope that presents a true picture of electronic events would be satisfactory. Likewise, simple multimeters, etc., are usable. For certification and return to service, however, only prescribed laboratory test equipment may be used.

Unlike the test equipment used for checking household radio and tv sets, most aviation test equipment is built to laboratory standards. The reason, of course, is that the accuracy and dependability of aviation electronics equipment is often vital to the safety of the airplane and its passengers; hence, tests to much higher standards are required. The tv repairman, for example, rarely if ever runs laboratory-type sensitivity tests; yet each VOR receiver certified for return service from an airline shop has had such a test.

The FAA requires that the test equipment used for a given test be the type specified by the manufacturer (or its equivalent). Tolerances on test-equipment performance are determined by those of the equipment under test. The management of the certified repair station is responsible for calibration of the test equipment in the shop. In general, the methods of testing are recommended by the manufacturer. It is the responsibility of the repair-station management to utilize the proper test methods. The specific test equipment chosen for a repair station is based on the rating sought. Besides the specialized test equipment, a conventional multimeter (volt-ohm-milliammeter or electronic voltmeter), frequency meter, and oscilloscope are required. For a Class-3 repair station, a triggered-sweep oscilloscope will be required. The equipment should be of standard quality.

Avionics shops may be classified according to the type of aircraft serviced. In general, this would be:

Single-engine aircraft.
Light, twin-engine, IFR-rated aircraft.
Jets and turboprops.

The investment required to service jets and turboprops, obviously, is the greater. The type of avionics equipment to be serviced may be listed in order of shop investment priority as follows:

Communications.
Navigation: VOR and Localizer.
ADF.
Marker Receivers.
Glide-Slope Receivers.
DME.
Transponders.
Radar.
Autopilots and Flight Directors.

Ramp Testing

Specialized test equipment is that intended for a singular purpose such as testing VOR systems. Specialized test equipment for avionics may be classified as either "ramp" or "bench." Ramp test equipment is used to provide a quick system check in or near the airplane.

The Tel-Instrument Electronics Corporation Type T-30/T-30A VOR/ILS ramp tester shown in Fig. 15-1 is designed for one-man operation to check VOR, localizer, glide-slope, and marker-beacon receivers. The rf signals are directly radiated to the aircraft antenna. The units contained in the group are operated from internal, 9-volt

Fig. 15-1. The TIC Type T-30/T-30A VOR/ILS ramp tester. (*Courtesy Tel-Instrument Electronics Corp.*)

alkaline batteries. Such equipment is now becoming standard with several operators and airlines as an economical safety check that often eliminates flight tests.

Another example of a ramp tester is the Collins 479U-1 signal generator shown in Fig. 15-2. This instrument quickly checks both VOR and glide-slope signals at predetermined frequencies. Powered from 13.3 or 26.5 volts dc, it may be attached directly to the equipment

Fig. 15-2. The Collins 479U-1 signal generator used as a ramp tester for VOR and slide-slope equipment. (*Courtesy Collins Div., Rockwell International*)

under test, or an antenna may be used to transmit the signal. Positioning the main control selects the various signals to be generated.

Signal Generators

Any of the less-expensive signal generators may be used to provide a signal for purely troubleshooting purposes, but are usually not adequate for quantitative tests. For testing ADF and hf equipment, the General Radio Type 1001-A standard signal generator is often specified. It operates from 5 to 50 kHz, providing up to 80% modulation with a 400-Hz internally generated signal. Provision is made for an external modulation signal. Output voltage can be varied from 0.1 microvolt to 200 millivolts. A full 2-volt signal can also be provided by bypassing the attenuator circuit. This instrument is very accurate, and its outputs are metered.

For vhf work, the Boonton 211-A signal generator is usually specified. It covers the range from 88 to 140 MHz and has an output of 0.1 microvolt to 200 millivolts into a 50-ohm load. Internal 400- or 1000-Hz modulating sources are provided, with modulation adjustable from zero to 100% in two ranges. A crystal oscillator with two ranges (110.1 and 114.9 MHz) is provided for calibration of a variable-frequency oscillator that can be tuned throughout the range previously noted.

A signal generator for specialized application is the Tel-Instrument T-15A. The T-15A is a crystal-controlled uhf signal generator designed specifically for testing of ATC transponders and DME interrogators. When used with a suitable pulse modulator and oscilloscope, this instrument provides complete test facilities. It consists of a crystal-controlled signal generator, high-power dummy load, variable piston attenuator, and rf detector. An additional feature is a provision to inject a crystal-controlled 1090-MHz cw signal into the detector circuit to provide a visual zero beat of an ATC transponder transmitter.

The T-15A covers 11 specific DME frequencies, chosen in such a manner as to permit the testing of all receiving crystals in a DME interrogator. This is possible since DME interrogators operate with one of 11 "MHz" crystals and with one of 10 "tenth-MHz" crystals, summed in a mixer, to provide any of the 100 vhf channels.

A simple low-cost backup vhf signal generator to the primary shop standard is the Radio Systems Technology Model RST-721 Communications Test Set (Fig. 15-3). This unit provides a single crystal-controlled 122.8-MHz signal modulated with either a 400-Hz tone for receiver test, or modulated by an external microphone. Microphone audio level is also read on the front-panel meter, and a panel lamp indicates proper microphone key operation.

The rf power output in two ranges (3 W, 10 W), audio speaker power (3 W, 10 W), headphone voltage (3 volts rms), vswr (50:1),

Fig. 15-3. The Radio Systems Technology Model RST-721 Communications Test Set. (*Courtesy Radio Systems Technology*)

and RST-601 power-supply voltage and current may be measured on the front-panel meter.

If vswr is not being measured, the vswr connector is a handy source of attenuated (−20 dB) transmitter signals for use with external low-level counters or other measurement devices.

A 400-Hz audio signal is available on the front panel, adjustable from zero to 2.5 volts rms. A pin on the rear-panel connector is reserved for oscilloscope monitoring of the transmitter audio.

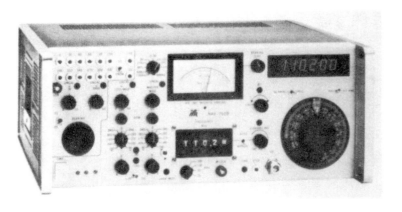

Fig. 15-4. The IFR, Inc., Model NAV-750B is a bench unit capable of complete performance analysis of VOR and ILS equipment. (*Courtesy IFR, Inc.*)

Rear-panel antenna jacks are provided for use of this test set as a portable ramp tester. A radiated signal from the plug-in antenna allows on-board test of the communications system. Portable operation from an internal 9-volt battery or bench operation from a 12-volt supply is switch-selectable from the front panel.

The IFR, Inc., line of test equipment features the NAV-750 series test sets for testing VOR and ILS equipment. These units feature modern design and ease of use combined with versatility.

The NAV-750 series bench test sets are precision simulators of VOR, ILS (localizer and glide slope), and communication ground stations. Simulation of the marker-beacon ground station with selectable tones for outer, middle, and inner marker beacons is included in the NAV-750B version (Fig. 15-4). The units can remotely channel a receiver using the ARINC two-out-of-five select system or parallel BCD code.

The VOR section provides push-button bearing selection for each 30° of azimuth as well as two push buttons which increase or decrease the bearing in 10° increments. The bearing may also be set in

0.01° or 0.05° increments with the bearing control. All VOR rf output levels may be varied with a calibrated output attenuator. The localizer and glide-slope sections use precision 90- and 150-Hz tone generators to provide accurately mixed tones for on-course and specific off-course signals at levels that may be varied with a calibrated output attenuator.

Communication receiver frequencies from 118.000 to 156.000 MHz are provided in 25-kHz steps. A 1020-Hz tone may be added. The rf output level may be varied with a calibrated output attenuator. The rf generator is crystal controlled and phase-locked in 25-kHz steps. Automatic frequency stepping at a variable rate is provided, selectable in 25-, 50-, 100-, or 200-kHz increments.

Modulating signals are provided on the rear panel for troubleshooting and special tests. For example, when operating in the VOR mode, the sum-of-tones output jack provides all tones necessary for checking VOR receiver bearing circuits without using the rf portion of the receiver. Individual tones are available at other jacks for similar purposes.

The NAV-750 automatically produces a standard VOR test signal, variable in rf level from −120 dBm to −6 dBm, whenever a VOR frequency is selected. The 30-Hz variable-tone percentage of modulation may be varied from 0% to 60% with the 30-Hz tone modulation level control. The 9960-Hz tone (frequency modulated with the 30-Hz reference tone) percentage of modulation can be varied in the same manner with the 9960-Hz tone modulation control. The 1020-Hz Ident tone is normally off in VOR and may be added by rotating the 1020-Hz modulation level control clockwise. The master modulation control will vary all tone percent modulation levels from 0% to 60%. VOR bearings may be selected using the 12 bearing push buttons, the +10° and −10° bearing push-button switches, or the variable bearing control. The bearing is displayed when the BEARING-FREQ. select switch is in the BEARING position.

Complete performance tests may be run without any other equipment. The technical manual provided with the NAV-750 series contains a complete description of controls and procedures for performance checks. In cases where a manufacturer specifies the NAV-750 for performance tests, these may be followed in satisfaction of any mandatory requirement. In any case, the user may devise various techniques for bench testing and troubleshooting as required.

When making performance tests, many shops use printed forms containing step-by-step procedures with a blank space for inserting actual measurements. The form may be general, or devised to fit a given make and model of Comm/Nav equipment. A form may contain actual step-by-step procedures, as in the following extract from the NAV-750 technical manual:

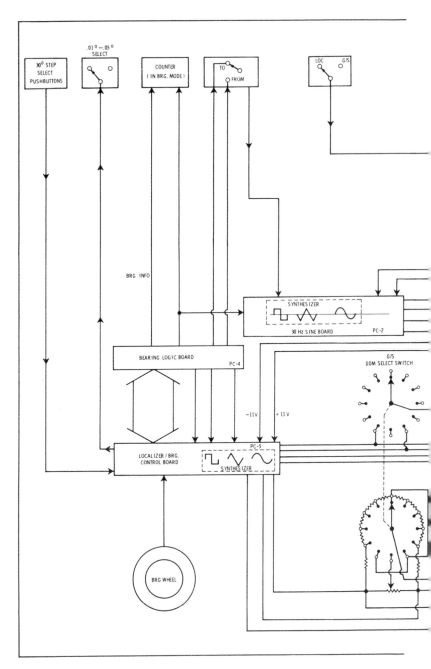

Fig. 15-5. Block diagram of the VOR section of the IFR, Inc.,

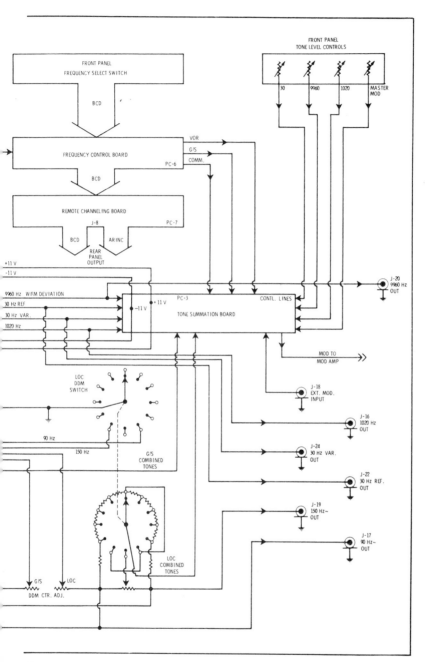

NAV-750 series. (*Courtesy IFR, Inc.*)

16. Rotate variable BEARING control clockwise until display is 12.00°.
17. Press push button. Note centering error and release push button.
18. Record centering error, due to tone phase shift, in pointer deflection or μA _____.
19. Rotate variable BEARING control counterclockwise until display is 348.00°.
20. Press ϕ push button. Note centering error and release ϕ push button.
21. Record centering error, due to tone phase shift, in pointer deflection or μA _____.

Another method of recording test results is to tabulate the measured performance limits in a handy form which may be filed and also supplied to the equipment owner when a unit is returned to service.

A block diagram of the VOR section of the NAV-750 is shown in Fig. 15-5. The VOR section is located mainly on PC-4 and PC-5. The VOR bearing is selected on these two boards. The bearing oscillator (2.16-MHz clock oscillator) is located on PC-4 and provides the reference signal to the VOR bearing counter system. Four tones are derived from frequencies obtained from the bearing counter system —the 30-Hz reference tone, 30-Hz variable tone, 90-Hz tone, and 150-Hz tone. The two 30-Hz tone signals are applied to PC-2 where the square waves are converted to sine waveforms combined with a 9960-Hz signal, and fed to PC-3.

The VOR tones are applied on PC-3 to controls that vary the modulation level. PC-3 also switches all modulation tones as determined by the output rf selection. The tones are then summed and sent to the modulator amplifier assembly where they modulate the rf output signal. The VOR bearing selected is displayed on a front-panel counter display. An RF DEMOD output is used to check the accuracy of the bearing at 90° and 270° by circuitry on PC-1.

ADF TESTING

A review of the principles of a loop antenna described in Chapter 7 reveals that the loop antenna operates through the effect of a signal arriving in space from a fixed radiation source. This means that no true accuracy test of an ADF receiver can be made unless the characteristics of such a signal can be simulated under controlled conditions. A radiation field must be set up in a controlled environment, and the ADF antenna must be in this environment for the test. The environment must be completely sterile of any extraneous radiation

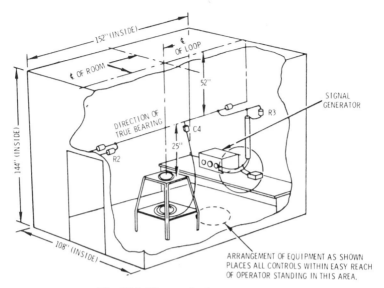

Fig. 15-6. The standard screen room.

fields. A number of other tests also require a sterile radiation environment. For these reasons, a "screen room" is often part of a well-equipped shop.

A standard screen room is constructed as shown in Fig. 15-6. If a screen room with dimensions other than those given in the illustration is required, new calculations must be made.[*]

The frame and floor are made of wood, and the room is completely enclosed with copper screen inside and out. The inner and outer screens should be connected together and grounded at one point only, although all screen joints must be soldered. When the door is closed, the door shields must securely contact the surrounding screen throughout its periphery. Radio-frequency filters must be used on all external power leads brought into the screen room. Line filters should be on the outside of the screen room and be bonded to the test room at that point where inner and outer shields connect. The output leads of the filter must be enclosed in conduit that is well bonded to the shield within the room.

For ADF testing, a calibrated transmission line, terminated in its characteristic impedance, is fitted overhead in the screen room. When fed from a standard signal generator, this line produces a field of known, controllable strength that can be used to induce currents in an ADF loop. The overhead transmission line is fastened to insu-

---

[*]See "Generation of Standard Fields in Shielded Enclosures," Fred Haber, *Proceedings of the IRE*, November, 1954.

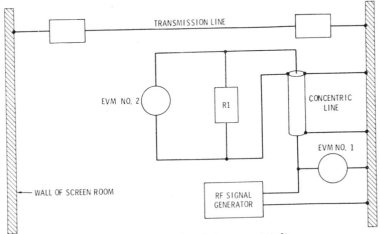

Fig. 15-7. Termination of the concentric line.

lators approximately four inches from the wall and stretched tightly. The line is fed from a signal generator, through a low-capacitance concentric line that is matched in three steps, as illustrated in Figs. 15-7, 15-8, and 15-9.

Termination of the Concentric Line

With reference to Fig. 15-7, disconnect the concentric line from the transmission line. Vary the value of R1 to obtain equal voltages at evm's Nos. 1 and 2 at a frequency having a wavelength four times

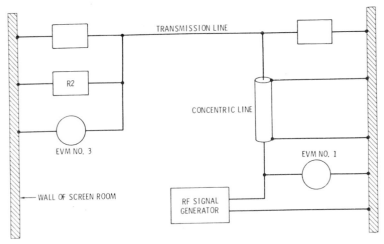

Fig. 15-8. Termination of the transmission line.

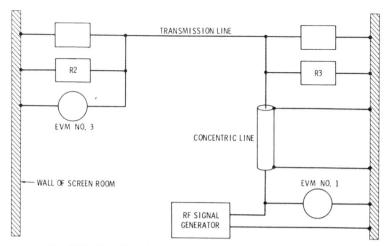

Fig. 15-9. Matching the concentric line to the transmission line.

the length of the concentric line. Check relative readings of the evm's at frequencies ranging from 50 kHz to 3 MHz. The readings on the evm's should be equal at all frequencies below that previously designated when the correct value for R1 was determined.

Termination of the Transmission Line

With reference to Fig. 15-8, remove R1 and connect the concentric line to the transmission line. Set the signal generator to that frequency at which the system appears one-fourth wavelength long.

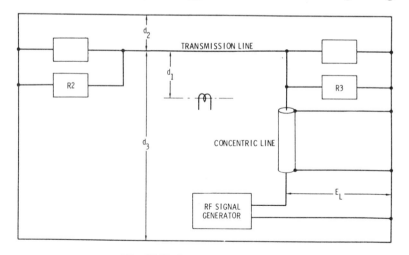

Fig. 15-10. Screen-room constants.

Vary the value of R2 for equal readings on both evm's Nos. 1 and 3, and check at lower frequencies as in the previous step. Measure the resistance of R2 and install it permanently.

Matching Concentric Line to Transmission Line

With reference to Fig. 15-9, compute the value of R3 from the following:

$$R3 = \frac{R1 \times R2}{R2 - R1}$$

Connect R3 into the circuit, and check the readings of evm's Nos. 1 and 3 over the range of 50 kHz to 3 MHz as before. Both evm readings should be equal.

This arrangement may now be used for establishing accurately known fields. With reference to Fig. 15-10, the field strength may be calculated as follows:

$$E_a = E_c + E_f - E_u$$

where,

E_a is the field strength at a given distance,
E_c is the field strength due to the current in the line,
E_f is the field strength due to the current in the floor,
E_u is the field strength due to the current in the ceiling.

The values of E_c, E_f, and E_u can be found from the following:

$$E_c = \frac{2.36 \times 10^3}{d_1}\left(\frac{E_L}{Z_L}\right)$$

$$E_f = \frac{2.36 \times 10^3}{2d_3 - d_1}\left(\frac{E_L}{Z_L}\right)$$

$$E_u = \frac{2.36 \times 10^3}{2d_2 + d_1}\left(\frac{E_L}{Z_L}\right)$$

where,

E_L is the line input voltage from the signal generator,
Z_L is the characteristic impedance of the transmission line and is equal to R2.

The attenuation constant (K) of the screen room is equal to:

$$\frac{E_L}{E_f}$$

Thus, if K is equal to 10, a signal-generator setting of 600 microvolts will produce a field strength of 60 microvolts at distance d_1. In practice, ADF loops are always tested at the same position in the screen room.

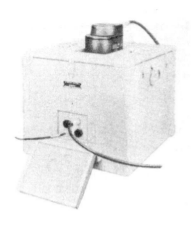

Fig. 15-11. The Transidyne Type G1 portable test shield. (*Courtesy Transidyne-General Corp.*)

It is also possible to test ADF receivers without a screen room by employing special test equipment. The Transidyne Type G1 portable test shield (Fig. 15-11) can substitute for a calibrated screen room. The loop under test is mounted inside the device, which is in the form of a copper-lined "rf weather-stripped" box. Inside the box, a transmission line, terminated in its characteristic impedance, radiates and thereby provides a calibrated radiation field. With a signal generator of 50-ohms internal impedance, the test shield has an at-

Fig. 15-12. The Collins 477U-1 Loop Simulator may be substituted for the loop and calibrated screen room during ADF tests. (*Courtesy Collins Div., Rockwell International*)

Fig. 15-13. The Carter Model CES-116A ADF Signal Simulator.
(*Courtesy Tel-Instrument Electronics Corp.*)

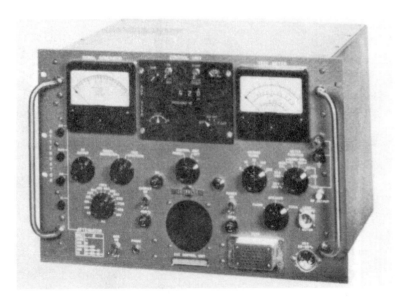

Fig. 15-14. The Collins 477V-1 ADF signal generator operates in conjunction with signal simulators such as shown in Figs. 15-12 and 15-13. (*Courtesy Collins Div., Rockwell International*)

tenuation constant of 0.2; that is, a 1-volt output from the signal generator will develop 200,000 microvolts per meter. Sense-antenna voltages equivalent to 100- and 150-pF antennas, and with effective heights variable from 0.05 to 0.5 meter, are provided, as well as sense antennas of 25 and 50 pF.

Another method of testing ADF receiver performance without a screen room is to use a loop simulator like the Collins 477U-1 shown in Fig. 15-12, or the Carter Model CES-116A ADF Signal Simulator shown in Fig. 15-13. Such simulators, used in conjunction with an ADF signal generator such as the Collins 477V-1 shown in Fig. 15-14, provide simulated antenna signals that are identical to signals arriving from space and received on an ADF antenna system.

16

Aircraft
Installations

All of the "black boxes" discussed so far must be suitably joined together to form the "electronic package." In new airplanes, the electronic equipment is usually installed at the factory, but new electronic systems are often installed in older airplanes as well. The first DC-3 airliners were built in the early 1930s and those that remain can still operate in many types of service. Yet they would not be permitted to fly over the airways with their original electronic equipment. The rapid advances over the past few years have, therefore, resulted in many retrofit programs.

Electronic installations are closely keyed to economic considerations. All-weather capabilities are most desirable, of course, because they permit full utilization of an airplane. Yet the cost of a complete all-weather electronic package easily exceeds that of many airplanes. Electronic manufacturers compete to produce equipment that can be combined to fit into each cost classification in a way that will provide maximum utility for the price. This puts the aircraft owner or chief pilot into the same position as the hi-fi enthusiast who chooses different pieces of equipment from the various product lines to form what he thinks will be the ideal combination for his purpose.

In addition to the cost, the weight, space, and power consumption are very important. Space and weight are at a premium in an airplane, and although the loss of a few pounds in payload may seem of slight importance in a single flight, it can become an appreciable burden over the useful life of the airplane. An electronic installation must allow easy access for inspection, maintenance, and exchange of units. Interconnection, shielding, and bonding is an art in itself. For fleet operators, interchangeability is of prime importance. The airlines, for example, have stipulated standardized equipment that can be interchanged with units of a different manufacturer.

CLASSES AND CATEGORIES OF AIRPLANES AND EQUIPMENT

An airplane is designed around its power plant. The range of thrust available to move the airplane through the air and the altitudes at which the thrust is available determine the ultimate cost of the airplane and its performance regime.

The following system of classifying airplanes is used by standardization groups, such as the RTCA:

Class X Aircraft
Class X includes aircraft with a maximum ceiling of 20,000 feet (6000 meters). This includes helicopters, light aircraft, piston-engined executive aircraft, and, in general, all unpressurized aircraft. Typically, these aircraft will use outside air for cooling and will have a simple heating system using waste engine heat.

Class Y Aircraft
Class Y includes most passenger and cargo aircraft having ceilings of 50,000 feet (15,000 meters).

Class Z Aircraft
Class Z includes all supersonic aircraft.

An airplane that flies at high altitudes will be moving through air at very low temperature and air pressure. Equipment that is installed outside of a heated pressurized area must be capable of withstanding this environment. There are other factors, also, which determine the operating environment of electronic equipment. These are categorized as follows:

Category A1
Equipment intended for installation in controlled temperature and pressurized locations in aircraft in which the pressures are no lower than that which is equivalent to an altitude of 15,000 feet (4500 meters) mean sea level (MSL).

Category A2
Equipment intended for installation in partially controlled temperature and pressurized locations in aircraft in which the

pressures are no lower than that which is equivalent to an altitude of 15,000 feet (4500 meters) MSL.

Category B1

Equipment intended for installation in nonpressurized and controlled temperature locations in Class X aircraft which operate at altitudes up to 20,000 feet (6000 meters) MSL.

Category B2

Equipment intended for installation in nonpressurized and noncontrolled temperature locations in Class X aircraft which operate at altitudes up to 20,000 feet (6000 meters) MSL.

Category B3

Equipment intended for installation in the power compartment of Class X aircraft which operate at altitudes up to 20,000 feet (6000 meters) MSL.

Category C1

Equipment intended for installation in nonpressurized and controlled temperature locations in Class Y aircraft which operate at altitudes up to 35,000 feet (10,500 meters) MSL.

Category C2

Equipment intended for installation in nonpressurized and noncontrolled temperature locations in Class Y aircraft which operate at altitudes up to 35,000 feet (10,500 meters) MSL.

Category C3

Equipment intended for installation in the power plant compartment of Class Y aircraft which operate at altitudes up to 35,000 feet (10,500 meters) MSL.

Category D1

Equipment intended for installation in nonpressurized and controlled temperature locations in Class Y aircraft which operate at altitudes up to 50,000 feet (15,000 meters) MSL.

Category D2

Equipment intended for installation in nonpressurized and noncontrolled temperature locations in Class Y aircraft which operate at altitudes up to 50,000 feet (15,000 meters) MSL.

Category D3

Equipment intended for installation in the power plant compartment of Class Y aircraft which operate at altitudes up to 50,000 feet (15,000 meters) MSL.

Category E1

Equipment intended for installation in nonpressurized and noncontrolled temperature locations in Class Z aircraft which operate at altitudes up to 70,000 feet (21,000 meters) MSL.

Category E2

Equipment intended for installation in the power plant compartment of Class Z aircraft which operate at altitudes up to 70,000 feet (21,000 meters) MSL.

EQUIPMENT STANDARDS

Part 37 of the Federal Air Regulations lists those devices for which Technical Standards Orders (TSOs) have been issued. Not all equipment need meet TSOs in order to be used in general aircraft, although air-carrier airplanes must have almost all on-board equipment TSOd. All transponders must meet TSOs.

Sometimes a TSO has not been established for a device or article, or if a TSO exists, the FAA may only require *approval*, rather than *qualification* under a specific TSO. In this case, materials, parts, processes, and appliances that require approval may be approved under a Parts Manufacturer Approval, in conjunction with type certification procedures for a product, or in any other manner approved by the FAA. Some examples of articles that must be approved, but not necessarily TSOd, are seat belts, aircraft position lights, anticollision light systems, and DME when required for operations at and above 24,000 feet (7200 meters) MSL.

The RTCA writes tests for electronic equipment which are later adapted as TSOs. These tests are contained in Document No. DO-160, Feb. 28, 1975, which may be obtained from the Radio Technical Commission for Aeronautics, Suite 655, 1717 H Street, N.W., Washington, DC 20006. Fourteen environmental tests have been established. These can be identified on the nameplate of electronic equipment. The symbols "DO-160 Env. Cat." are followed by a combination of 1–15 letters and numbers that identify the environmental categories as follows:

1. Temperature and altitude test (2 spaces minimum).
2. Humidity test.
3. Vibration test.
4. Explosion test.
5. Waterproofness test.
6. Hydraulic fluid test.
7. Sand dust test.
8. Fungus resistance test.
9. Salt spray test.
10. Magnetic effect test.
11. Power input test.
12. Voltage spikes test.
13. Audio frequency conducted susceptibility test.

14. Electromagnetic compatibility test.
    Induced signal susceptibility test.
    Radio frequency susceptibility test.
    Emission of radio frequency energy test.

A typical equipment nameplate identification is as follows:

DO-160 Env. Cat. A2AKXXXXXXXAAAAA

A manufacturer may qualify equipment to more than one category for a particular environmental test. If one category is more stringent, only that category need be identified. In cases such as temperature/altitude or vibration, where the test requirements for various categories are different but not necessarily more severe, more than one category will be marked on the nameplate; for example, the following identification is identical to the foregoing example, but the equipment has been qualified to temperature and altitude test categories A1 and D2 and vibration categories M, N, and O:

DO-160 Env. Cat./AID2/A/MNO/XXXXXXAAAAA

In the vibration test, equipment may be qualified for one category without shock mounts and to another with shock mounts. The differentiation is shown by listing test categories without shock mounts above the line and those with shock mounts below the line. For example, the following nameplate identification is identical to the previous example, except that the equipment has been qualified to vibration categories M, N, and O without shock mounts and to category J with shock mounts:

$$\text{DO-160 Env. Cat./AID2/A/}\frac{\text{MNO}}{\text{J}}\text{XXXXXXAAAAA}$$

The letters refer to "categories" of the specified test as they are listed in RTCA Document DO-160.

EQUIPMENT LAYOUTS

The heavier, higher-powered, and more-sensitive equipment specified by the airlines (ARINC) has only the indicator and control devices mounted in the cockpit; the actual "black boxes" are mounted in a special area, usually on an "electronics rack." Airline equipment based upon ARINC standards employs such equipment racks located in an assigned compartment. A defective unit may be quickly exchanged during an airport stopover. Defective or suspect units are then sent to a central repair shop for repair and test. Fig. 16-1 shows a rack containing dual IFR equipment. Designers of equipment for general aviation, who have more freedom

Fig. 16-1. A typical airline installation is illustrated by the Wilcox Model 723 package. (*Courtesy Wilcox Electric Co.*)

and are not bound to airline specifications, usually prefer to mount as much of the equipment as possible on the instrument panel.

The simplest electronic requirement for VFR flight would be for air-to-ground vhf communication, and could be met by the installation of a simple panel-mounted transceiver. A second minimum requirement would be a VOR receiver.

The demands of IFR flight become more severe. An ADF is usually added for IFR operations, and two VOR receivers greatly facilitate quick and accurate VOR fixes. The ILS is added if it is not desired

349

to depend solely on PAR. As we proceed up the scale, for consistent all-weather operations, two communications, two VOR, and two ADF systems are used. The advantages of such duality are that one system can be checked against the other, and failure of one system will not result in the declaration of emergency.

General aviation installations fall into three classifications. First, there are low-cost, panel-mounted systems that do not meet TSO. Second, there is a middle class that meets TSO. Third is the airline class of equipment. The term "one and one-half," often used in general aviation circles, applies to a system employing one transmitter, a communications receiver, and another receiver for navigation. The advantage of the "1 + ½" system is that the pilot may communicate with the ground without loss of VOR information. The time spent in changing a single receiver from communication to VOR station and vice versa is prohibitive. Hence the term "1½."

From this point, more sophisticated systems are built up. The use of a special central switching panel is desirable for switching over microphone and speakers, controlling volume, etc.

Heat is an important factor affecting the life of electronic equipment. Heat causes insulating materials to shrink and crack. Glass-to-metal seals become defective. Installations must be made to allow maximum heat dissipation. The effect of direct sunlight must be taken into consideration, even in flight. In larger installations, blowers must be utilized.

The instrument panel of a twin-engine Piper *Navaho Chieftan* is shown in Fig. 16-2. The area in the center above the throttle quadrant contains vhf Comm/Nav, ADF, DME, transponder, and weather radar. The pilot-in-command on the left has a flight director and radio magnetic indicator. The right seat has a gyrohorizon only, with gyrocompass below.

The choice of equipment depends on the type of aircraft and the use put upon it. The Narco "Basic VFR #1" equipment package, shown in Fig. 16-3, would provide a system for flying in noncongested areas. The Mark 8 NAV/COM unit will provide basic vhf communications, with VOR/LOC navigation. The ADF-31A automatic direction finder will give bearings to lf beacons and commercial broadcast stations.

For VFR flying where higher cost is permissible, the Narco "Standard VFR #2" package (Fig. 16-4) is an example. This combination consists of a Narco Mark 12B NAV/COM transceiver with 360 channels for communication and 100 channels for navigation, a VOA-8 VOR/ILS localizer, an ADF-31A automatic direction finder, and a UDI-4 DME receiver. The package allows maximum cross-country VFR capability. For operating in congested airports, a Narco AT6-A transponder would be a desirable addition.

Fig. 16-2. Instrument layout in a Piper *Navaho Chieftan*. (*Courtesy Piper Aircraft Corp.*)

Fig. 16-3. The NARCO "Basic VFR #1" equipment package. (*Courtesy NARCO Avionics, a Division of NARCO Scientific Industries, Inc.*)

Fig. 16-4. The NARCO "Standard VFR #2" equipment package. (*Courtesy NARCO Avionics, a Division of NARCO Scientific Industries, Inc.*)

For IFR flight, the Narco "Primary IFR #1" package (Fig. 16-5) is an example of a minimum requirement. This package permits radar monitoring with the transponder, and advisory service. The Mark 12B covers all frequencies, with a Mark 8 for identifying intersections and as a backup. A VOA-50M VOR/ILS localizer, with glide-slope cross-pointer, teams with the Mark 12B. The ADF-31A provides bearings to beacons and commercial stations. The UDI-4 DME provides quick navigation fixes. The AT6-A transponder provides radar identification. The Narco VP22 master selector-switch system switches the audio system to the various NAV/COM systems.

Fig. 16-5. The NARCO "Primary IFR #1" equipment package. (*Courtesy NARCO Avionics, a Division of NARCO Scientific Industries, Inc.*)

Additional versatility is provided by the Narco "Full IFR #2" package shown in Fig. 16-6. A system such as this will permit operation for all-weather flying. The package contains two Mark 16 NAV/COM units with 360 channels for communication and 100 channels for navigation, two VOA-50M VOR/ILS localizers with glide-slope indicator (and remote receiver), and marker lights for outer, middle, and inner markers to permit low-ceiling (CAT II/ILS) approaches. The UDI-4 DME/GSI in one unit, or UDI-2ARD DME/GSI remote system, which gives distance measuring up to 150 miles and ground speed to 300 knots, provides use of all VORTAC facilities. The AT6-A transponder, with altitude reporting capability, provides recognition-identification on the air traffic controller's radar screen. The ADF-31A automatic direction finder covers low-frequency facilities and has a built-in aural marker-beacon receiver. The VP22 master selector-switch system puts all avionics switches in one row.

Fig. 16-6. The NARCO "Full IFR #2" equipment package. (*Courtesy NARCO Avionics, a Division of NARCO Scientific Industries, Inc.*)

AUDIO SYSTEMS

The audio systems of aircraft deserve special attention. Many aspects of audio design are beyond the control of the electronic manufacturer, and are left to the installing agency. The pilot must be able to clearly hear the output of his several receivers. He must also be provided with a means of easily switching from one system to another.

One audio system is the King Isolation Amplifier Model KA-25C. This system makes possible the combination of all systems into a common speaker with low distortion and flat frequency response. An additional feature is the use of a "ramp hailer," which is an external public-address type speaker located in the front wheel well or nose, which allows the direction of flight-line personnel from the cockpit.

A microphone selector switch with three positions can be installed to provide microphone control circuits to either No. 1 or No. 2 vhf equipment, or to a third position which then properly routes the microphone output to the external ramp speaker or to a cabin speaker in aircraft utilizing a separated cockpit and cabin area.

An electronic muting circuit is provided to automatically isolate the output of all receivers from the isolation amplifier whenever the microphone button is pressed. This feature eliminates the possibility of audio feedback in the cockpit.

A series-type, solid-state voltage regulator is used to supply power for the low-level audio stages of the isolation amplifier. Consequently, no wiring changes are necessary when switching from 14 to 28 volts. The voltage regulator also acts as a deterrent to alternator noise and generator ripple.

A schematic of the KA-25C is shown in Fig. 16-7. The audio signals from the various receivers are coupled through isolation resistors R123 through R130, CR102, and C110. Diode CR102 is the audio muting diode that disconnects all receiver inputs from the isolation speaker amplifier during transmit or ramp-hailing operation. Note that resistor R117 is connected to the microphone key (terminal H), which is connected directly to the control contact of the pilot's microphone. When the microphone button is not depressed, the transmit control line is ungrounded, and diode CR102 is turned on by current from the regulated supply line through resistors R115, R117, R116, diode CR102, and resistor R118 to ground. Under this condition, diode CR102 is biased on, and audio signals are coupled to the base of Q105 through dc isolating capacitor C110. When the microphone button is depressed, the junction of R115 and R117 is grounded and the anode side of CR102 is reduced to ground potential as the voltage developed across C109 decays to zero. Capacitor C109 provides a transient filter to keep the sudden change in voltage from being transferred at an audio rate to the input of the isolation amplifier. To further ensure that CR102 will be cut off during transmit conditions, a slightly positive voltage is applied to the cathode by the connection of R119 to the regulated supply line.

Audio signals appearing at the base of Q105 are amplified and coupled from the collector to the base of driver transistor Q104 by

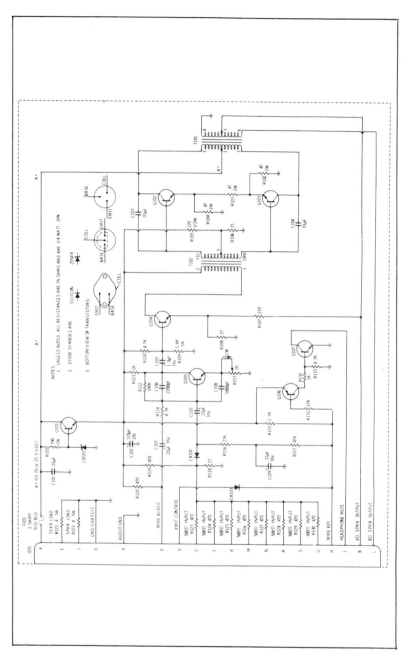

Fig. 16-7. Schematic of the King Model KA-25C audio amplifier.
(*Courtesy King Radio Corp.*)

capacitor C105. Driver amplifier Q104 is operated as a grounded-emitter amplifier, and the amplified signal appearing at the collector is coupled via transformer T102 to the bases of the Class-B push-pull stage comprising transistors Q102 and Q103. Bias supply for the final amplifier transistors is supplied to the center tap of the secondary of T102. The resistor network in the emitters of the amplifier provide balance and temperature compensation. Capacitors C103 and C104 provide high-frequency roll-off for the amplifier. The push-pull output of Q102 and Q103 is applied to the primary of output transformer T101. The secondary of T101 has taps at both the 4-ohm (terminal No. 5) and 8-ohm (terminal No. 6) terminations.

The last two stages of the amplifier are operating with inverse feedback to reduce distortion and to provide flat frequency response. This feedback is applied from terminal No. 5 of transformer T101 to the emitter of Q104 through resistor R107. The choice of coupling capacitors C110 and C105, and high-frequency roll-off capacitors C106, C103, and C104, are selected to provide a nearly flat frequency response from 300 Hz to 6000 Hz. The voltage-supply regulator, mostly comprised of Q101 and CR101, supplies closely regulated and highly filtered voltage to the first two amplifier stages of the isolation speaker amplifier. CR101 is a zener-type, voltage-regulator diode. This diode is operated in the reverse condition and shows a sharp increase in current at a selected voltage level. Thus, voltage at the base of Q101 is held nearly fixed even though the supply current through R101 should vary considerably. The zener diode and R101 are chosen such that after the supply voltage applied to the power connection (pin A) rises above 11 volts, the supply voltage may be raised to 30 volts without any significant increase in the voltage at the base of transistor Q101.

Due to the current gain of transistor Q101, the supply voltage to the collector may also vary considerably without material change in the emitter voltage from which the regulated supply energy is taken.

To provide an extremely low output impedance at higher frequencies, a tantalum capacitor, C102, is paralleled between the emitter and ground. It is because of the regulator that the isolation speaker amplifier is not susceptible to line-voltage variations, noise created by the aircraft alternator, and other disturbances. Unregulated line voltage is supplied to the collector circuit of the Class-B amplifier; however, because no further amplification takes place after this stage, and because of the noise-canceling effect of the inverse-feedback loop (terminal No. 5 of T101 to emitter of Q104), line transients still do not appear in the speaker output. The isolation speaker amplifier will provide 8-watts output with a 28-volt supply.

Undistorted output with a 14-volt supply is limited to 6 watts because of the lower voltage supplied to the Class-B amplifier.

OTHER ELECTRONIC INSTRUMENTATION

Electronic circuitry is often found within the instrumentation of an aircraft. This then becomes part of the avionic considerations of installation.

One example of this is the exhaust-gas temperature (EGT) instrument. The temperature of exhaust gas is directly related to the full air mixture. The correct carburetor mixture is usually that with the lowest exhaust-gas temperature, yet not over-rich (leaner mixtures run hotter).

Temperature is measured by a thermocouple placed in the exhaust stream. A thermocouple is essentially a pair of electrical conductors of dissimilar metals so joined as to produce a thermal emf when the junctions are at different temperatures. The conductors are joined at the end which is exposed to the temperature to be measured. The other ends are connected to an amplifier as shown in Fig. 16-8. The indicating instrument is a simple galvanometer. Calibration is easily accomplished by varying the amount of feedback, hence, gain.

Another instrument for monitoring fuel-air mixture is the exhaust-gas analyzer (EGA). The exhaust gas of a gasoline engine contains water vapor, carbon dioxide, carbon monoxide, oxygen, hydrogen, nitrogen and a small percentage of other gases. The fuel-air ratio determines the proportion of these gases. The principle of the EGA instrument is based upon the difference in heat conduction between hydrogen and carbon dioxide. Hydrogen conducts heat 12 times better than carbon dioxide.

In this instrument a small portion of exhaust gas is drawn off and allowed to cool. A heat sink is provided with a fixed amount of heat energy from an electric coil. The exhaust gas sample is then caused

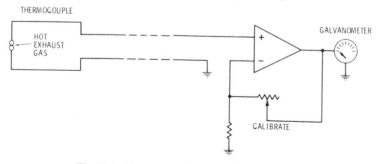

Fig. 16-8. Thermocouple connected to amplifier.

to cool the heat sink. An over-lean mixture produces a larger proportion of hydrogen, hence greater cooling. Usually, there are two identical heat sinks, one cooled by air, the other by the exhaust gas mixture. The temperature difference between them indicates the proportion of hydrogen. Thermocouples then supply a current in proportion to the heat sink temperatures, and a differential amplifier with its two inputs connected to the two thermocouples will reveal temperature difference.

Another instrumentation function involving electronic circuitry is in connection with fuel monitoring. On-board fuel and rate of fuel consumption are important flight parameters.

Less-expensive single-engine aircraft use float level instruments similar to automotive-type fuel gauges. The method is inadequate for larger tanks, however, and notoriously inaccurate. The electrical capacitance method is based upon two parallel plates immersed in the fuel. The fuel will have certain dielectric qualities differing from air, hence, the fuel level, acting as a dielectric, will vary the capacitance between the two plates. The capacitive reactance of the capacitor thus formed will be a measure of fuel level. In one such system, the fuel level units are concentric tubes electrically isolated from each other except for the dielectric. As the fuel level rises and falls, the dielectric varies. Capacitive reactance, X_c, is equal to:

$$X_c = \frac{1}{2\pi fC}$$

where,

X_c is the reactance in ohms,
f is the frequency in hertz,
C is the capacitance in farads.

We note from this relationship that as C grows larger, reactance becomes less. Likewise, as frequency increases, reactance becomes less. In practical systems, an oscillator is used to produce a current at a frequency on the order of 20 kHz. This frequency provides a low reactance, hence, larger magnitude of current change with fuel level. It is low enough in frequency, however, to preclude radiation problems, etc. A circuit for a capacitance-type fuel-monitoring meter is shown in Fig. 16-9.

There may be two or more of the capacitance pickups used in this system. This is to counter the effect of fuel surging from flight turbulence and to compensate for irregular fuel tank shapes. The various pickups are connected in parallel.

Fuel flow, usually in pounds per hour, can be measured in several ways. One method is to use mechanical vanes, such as those used in the pumps of service stations. These vanes can quite accurately reveal both rate of flow and total consumption. The vanes produce a

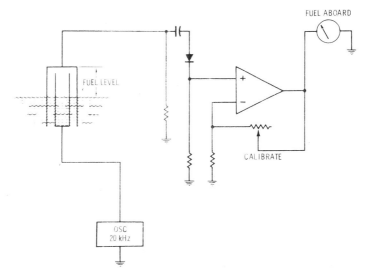

Fig. 16-9. Circuit for a capacitance-type fuel-monitoring meter.

velocity which is a function of fuel flow. Each revolution can be counted and will represent an amount of fuel. Circuitry is used to derive amount of fuel on board (departure capacity less that consumed) and rate of flow (usually in pounds per hour). Fuel management consists of judicious power settings versus fuel on board.

Instrumentation in larger aircraft usually features lightweight servo units employing 400-Hz power. The purpose of the servo often is only to position an indicating needle.

STANDARD SIZES AND FORM FACTORS

Considerable ingenuity is required if good installations are to be provided in the congested cockpits of today. As more and more electronic equipment was added to the cockpit, the need for standardization became more apparent. Pilots flying different models of airplanes will have less trouble in becoming familiar with the equipment if some degree of standardization is maintained. Moreover, cutting of panels is unnecessary when different electronic equipment is installed. Airplanes used for both military and civil operations should have the same panel facilities, to make equipment interchange easier. Electronic equipment control-panel dimensions are standardized by RTCA and are shown in Figs. 16-10 and 16-11.

Airline operators were the first to be faced with the necessity of standardized case sizes for radio equipment. In 1940, specifications defining the size of a standard Air Transport Radio (ATR) case

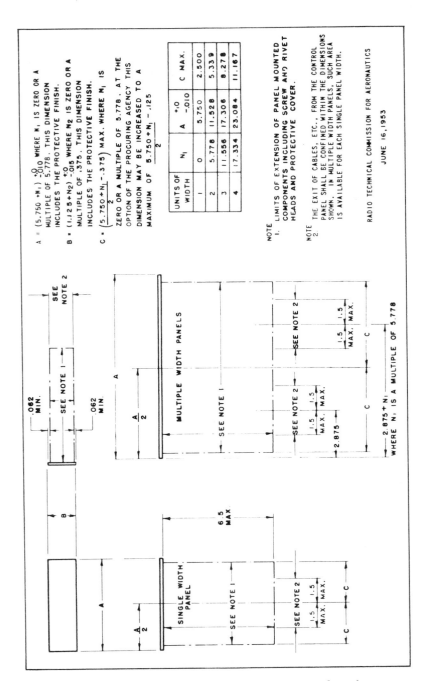

$A = (5.750 + N_1) ^{+0}_{-.010}$ WHERE N_1 IS ZERO OR A MULTIPLE OF 5.778. THIS DIMENSION INCLUDES THE PROTECTIVE FINISH.

$B = (1.125 + N_2) ^{+0}_{-.015}$ WHERE N_2 IS ZERO OR A MULTIPLE OF .375. THIS DIMENSION INCLUDES THE PROTECTIVE FINISH.

$C = \left(\dfrac{5.750 + N_1}{2} - .375\right)$ MAX. WHERE N_1 IS ZERO OR A MULTIPLE OF 5.778. AT THE OPTION OF THE PROCURING AGENCY THIS DIMENSION MAY BE INCREASED TO A MAXIMUM OF $\dfrac{5.750 + N_1}{2} - .125$

UNITS OF WIDTH	N_1	A $^{+0}_{-.010}$	C MAX.
1	0	5.750	2.500
2	5.778	11.528	5.339
3	11.556	17.306	8.278
4	17.334	23.084	11.167

NOTE 1. LIMITS OF EXTENSION OF PANEL MOUNTED COMPONENTS INCLUDING SCREW AND RIVET HEADS AND PROTECTIVE COVER.

NOTE 2. THE EXIT OF CABLES, ETC., FROM THE CONTROL PANEL SHALL BE CONFINED WITHIN THE DIMENSIONS SHOWN. IN MULTIPLE WIDTH PANELS, SUCH AREA IS AVAILABLE FOR EACH SINGLE PANEL WIDTH.

RADIO TECHNICAL COMMISSION FOR AERONAUTICS

JUNE 16, 1953

Fig. 16-10. Form factors for aircraft equipment control panels.

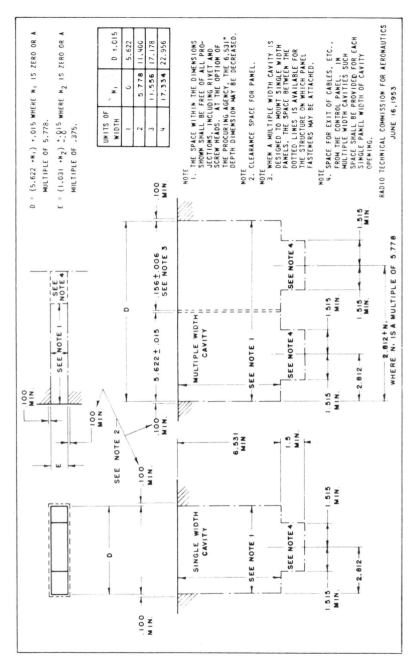

Fig. 16-11. Cavities for aircraft equipment control panels.

	Military Designation MIL-C-172	Approximate Volume cu. in.	(W) Width ±0.03125	(L) Length	(H) Height (Maximum)
	Short Quarter ATR	215	2.250	12.5625	7.625
	Long Quarter ATR	335	2.250	19.5625	7.625
	Short Three-Eighths ATR	340	3.5625	12.5625	7.625
	Long Three-Eighths ATR	530	3.5625	19.5625	7.625
A1-B	Short Half ATR	470	4.875	12.5625	7.625
A1-D	Long Half ATR	725	4.875	19.5625	7.625
	Short Three-Quarter ATR	720	7.50	12.5625	7.625
	Long Three-Quarter ATR	1120	7.50	19.5625	7.625
B1-D	One ATR	1510	10.125	19.5625	7.625
C1-D	One and One-Half ATR	2295	15.375	19.5625	7.625

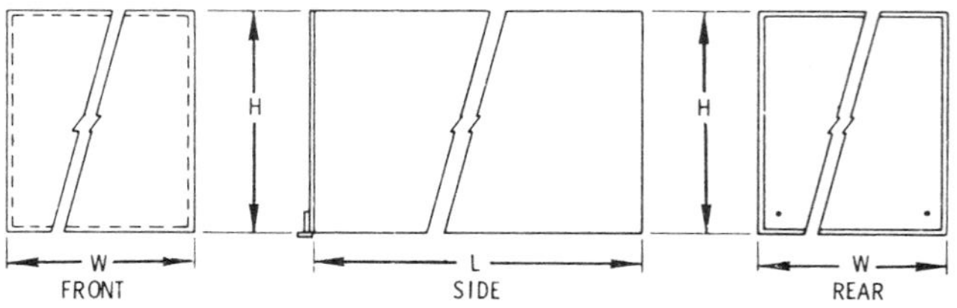

Fig. 16-12. Standard ATR case sizes.

Fig. 16-13. A standard ARINC shock mount.

for the electronic equipment of the Douglas DC-4, then under development, were drawn up by ARINC. There are ten standard ATR
sizes, as shown in Fig. 16-12. All leads from the equipment are
brought out to a specific plug in the rear of the case. The case itself
is mounted in a standard shock mount of the proper size. On the rear
of the shock mount is the mating plug, which fits into its counter-
part on the ATR case. The case is fastened securely into the shock
mount by two large finger screws, as shown in Fig. 16-13. This
arrangement greatly simplifies removal of equipment. As a further
step, ARINC specifies standard Cannon Type DPA plugs, and even
the pin numbers. One brand of ADF, for example, must therefore
be able to fit the same rack of another manufacturer. Although such
close standardization may discourage design initiative, the procedure certainly has freed the airlines from problems and thus contributed to their growth.

Most of the degenerative mechanical effects of an airplane on
electronic equipment are vibratory. The mounting for the ATR case,
although called a "shock mount," actually acts more like a vibration
damper. There is a great similarity between shock and vibration
mounting. Both utilize a resilient material held in shear and compression. Shock mounts can store more energy for a given deflection

363

than vibration mounts. The latter are designed so that the natural mechanical resonant frequency is lower than the lowest vibration anticipated. Unfortunately, a resilient mount designed for vibration resistance is not equally suited for shock resistance. Vibration mountings have such large deflections that a shock can easily cause "bottoming" unless the mount is only partially loaded.

In large installations, considerable heat is generated by the equipment on the electronic rack. In some units, internal blowers provide forced-air cooling. However, at the present time, integrated forced-air cooling is used on large airliners. Standard ARINC cooling is by means of openings that are located at the back of the ATR case and engage standard ducts on the electronic rack (Fig. 16-14). Air is sucked through the case into the duct, and then exhausted. As a rule of thumb, 30 pounds of air per 100 watts at a cabin altitude of 8000 feet will cause a temperature differential of 47°F. As a design criterion, 80° is used as the cabin temperature.

Electronic equipment must not depend on forced-air cooling, however. The purpose of external cooling, as presently envisioned, is to ensure longer life of equipment and to prevent dissipation of heat into the interior of the airplane.

Fig. 16-14. Two "One-ATR" units with ducted cooling in the lower navigation rack of a large airliner.

AIRCRAFT ELECTRICAL SYSTEMS

In place of a battery, the ignition systems of aircraft engines employ highly efficient dual magnetos. Early aircraft engines had no power takeoff for generators, and electrical requirements were scanty. Night flying introduced the need for navigation and cockpit lighting. Windmill generators were first used to charge the batteries, but were soon followed by power takeoffs for integral generators. Self-starting engines require larger batteries and generator capacities beginning at 150 watts. Light, single-engine private aircraft usually utilize 12-volt systems. The lower the voltage, the higher the current must be to produce the same amount of power; but high currents mean more losses in distribution unless heavier wiring is used. To decrease the weight of wiring, large aircraft utilize 24-volt dc systems.

Voltage values for the system are based on a nominal battery-terminal voltage of 2 volts per storage cell. In operation, the voltage of the system is determined by the battery and generator in parallel; and to allow charging current to flow, the generator must produce a voltage in excess of the "back voltage" of the battery. Aircraft equipment is therefore designed for 13.5 and 27.5 volts for 12- and 24-volt systems, respectively. Fluctuations of several volts can be expected between conditions of engine off with undercharged batteries, and engine at cruising power with fully charged batteries.

The electronic package of a light, twin-engine airplane may require only 10 amperes, but power requirements rise sharply with aircraft size. In transports, power is needed for lighting, in the galley, and for many other purposes. With such larger aircraft, several 28-volt generators with an average drain of 1500 to 2500 amperes are necessary. Special blast cooling allows lightweight generators to provide large amounts of power. Gasoline-powered portable generators must be used during ramp operations.

Fig. 16-15 shows the schematic of an electrical system using a carbon-pile voltage regulator to maintain a constant voltage. The charging rate depends on the discharge state of the batteries. The carbon pile consists of one or more stacks of carbon plates held in compression by a spring, in such a manner that minimum resistance is provided. As the batteries approach a fully charged condition, the voltage begins to rise. This activates the solenoid, which opposes the spring tension and allows the carbon stacks to become more loosely associated. As a result, their overall resistance increases and thereby reduces the current through the generator shunt windings. The lower field current reduces the generator voltage and thus provides a degenerative form of regulation. Adjustment of the rheostat sets the desired voltage for the circuit. Under some load conditions, the regulator may exhibit the undesirable tendency to "hunt." This is prevented by the compensating winding, which is designed to oppose any instantaneous reaction of the main winding. The reverse-current relay opens whenever generator voltage falls below a certain point, and thus prevents current from flowing back through the generator and discharging the batteries. The three possible adjustments to this voltage regulator are:

Voltage setting—Setting the current through the voltage-regulator solenoid.

Voltage sensitivity—Adjusting the spring tension on the carbon piles.

Voltage stability—Setting the current in the compensatory windings.

Adjustments are interdependent; changing one setting requires slight readjustment of the others.

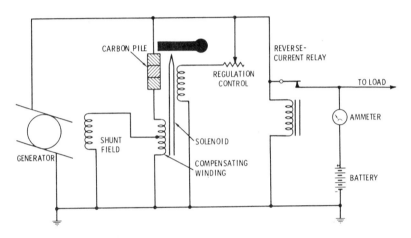

Fig. 16-15. Schematic of an aircraft battery-charging system using a carbon-pile voltage regulator.

The greatest power requirement, other than for starting the engines, is under flight conditions when the engines are set for cruising or climbing power. Storage batteries are therefore unnecessary for the bulk of the power requirements. More power per pound can be obtained from ac alternators. Furthermore, the use of alternators permits higher voltages and hence less copper losses. The frequency in hertz of an alternator is:

$$\frac{\text{Revolutions per Minute} \times N}{60}$$

where,

N is the number of poles.

If the alternator is connected directly to the engine shaft, the frequency will vary with engine rpm. Such a system is called "wild ac." Its major disadvantage becomes apparent when its effect on a reactive load is considered. Since the power factor can never be made constant, it is difficult to design efficient equipment for a "wild ac" system. Heating elements for wing deicing and other purposes can utilize "wild ac" more efficiently because the purely resistive nature of heater elements does not introduce power-factor problems. The newer generation of airliner employs constant-frequency alternators made possible by constant-speed drives, which take off power from the turbine shaft of a turbojet; and, by hydraulic means, hold the speed remarkably constant. These systems employ 400-Hz three-phase delta or wye configurations, using methods similar to those of ground power-distributor systems.

There is often a need for ac power at constant frequency. This form of power is required for servos and synchros. A frequency of 400 Hz has become the standard for avionic systems. This higher frequency results in the requirement for less iron, hence, weight in transformer cores and armatures. The reason for this is that inductive reactance, the ability to induce current, increases directly with frequency.

In most airplanes under the 12,500-pound class, batteries are the primary source of power. In order to provide ac, therefore, an inverter is used. Before the development of solid-state electronics, inverters were merely dc motors driving ac generators. Inverters now are "static," that is, require no moving parts. The efficiency of static

Fig. 16-16. This inverter is for use in airplanes equipped with 24-volt dc systems. (*Courtesy Avionic Instruments, Inc.*)

inverters is very high when compared to the older motor-alternator types. In the static type, a regenerative circuit operating at the desired frequency induces the alternating current into a winding for distribution within the airplane.

Fig. 16-16 shows the Avionic Instruments Model 1A1000 static inverter. This device meets the following specifications:

Input Voltage	
Nominal	28 V dc.
Normal Range	24–32 V dc.
Abnormal Range	18–37 V dc.
Input Current	
Nominal	48 A dc.
Maximum	80 A dc.
No Load	3 A dc.
Output Voltage	115/26 V ac ± 2.5 V typical.
Output Frequency	400 Hz ± 1%.
Output Distortion	4% maximum.
Output Power	1000 VA.
Power Factor Load	0.9 lead to 0.8 lag.
Waveform	Sine.
Overload	110% 2 hours.
	150% 5 minutes.
Short Circuit Current	14.0 A ac minimum 1 minute.
Efficiency	75% typical.
Weight	24 lb maximum.
Length	12.68 in.
Width	10.00 in.
Height	6.00 in.
Altitude	50,000 ft.
Temperature	−55°C to +71°C.

USA FAA TSO-C73, DO-138
MIL-E-5400 Class 2

If there is a need for three-phase power, three inverters of this type may be hooked together using a "phase-lock box" as shown in Fig. 16-17. The phase-lock box places each unit at the same frequency with phases 120° apart as in a standard three-phase system.

CABLES AND CONNECTORS

In the years before World War II, multiconductor cables were used for interconnection between radio equipment. The great variety of wiring combinations forced this method to be discontinued. Today, cables are made of bundles of wires suitable for a given installation. Except where a group of conductors must be shielded, the "wire-bundle" type of cable is now standard. Cables are fabricated by the agency that makes the installation. Suitable terminations are provided and continuity tests made, after which the cable is installed in the aircraft with suitable brackets or fasteners.

Insulated wire for aircraft cabling must meet strict specifications in regard to heat and abrasion resistance, for example. The insulation must be able to resist external heat, as well as heat produced

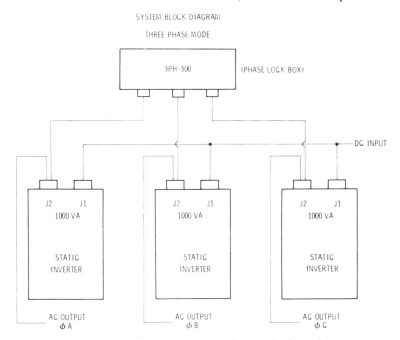

Fig. 16-17. Three inverters may be used to supply three-phase ac.

from within by excessive current through the conductor. Insulation is made from materials that have been subjected to *polymerization,* a process in which the molecules are enlarged to impart desirable qualities to the material. The diameter of wire is expressed as the "gauge." The higher the gauge, the less the cross-sectional area. The gauge to be used is determined by the amount of current the wire must carry. Table 16-1 shows the correct size for a given application.

Practically all plugs and sockets used with electronic equipment meet so-called "MIL" (military) standards. Plugs and receptacles are made in a wide variety of ways. One example is the simple standard plug connector shown in Fig. 16-18. The connector is affixed to the associated cable as follows:

Table 16-1. Wire Size Versus Current Rating for Single and Multiconductor Cables

Cable Size	Free Air Current Rating (amp)	Cables in Conduit or Bundles. Only three cables or 20% of the bundle, whichever is the greater, carrying maximum current simultaneously (amp)
20	11	7.5
18	16	10
16	22	13
14	32	17
12	41	23
10	55	33
8	73	46
6	101	60
4	135	80
2	181	100
1	211	125
0	245	150
00	283	175
000	328	200
0000	380	225

1. Install the cable bushing, cable clamp, rubber washer, metal washer, clamp adapter, and connector ring on the cable as shown.
2. Strip and tin the wires ¼ inch.
3. Solder the correct wire to the correct pin, using sleeving on each wire.

As soon as plugs have been installed, continuity should be checked. At the same time, pins are checked for correct wiring.

Connectors should be designed to resist moisture. One method is known as "potting," whereby plastic compounds are deposited around the pin connection and form a firm moistureproof bond. Potting compounds are usually one of the epoxy resins which must be mixed shortly before use. Hardening requires 6 to 24 hours and the plugs cannot thereafter be disassembled. A type of applicator is used to deposit the material firmly within the shell of the plug.

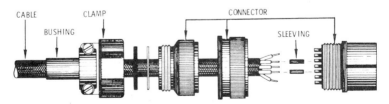

Fig. 16-18. A connector assembly.

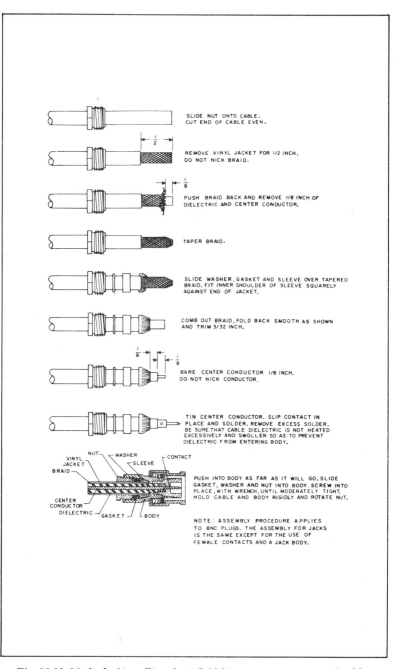

Fig. 16-19. Method of installing the UG-88/U connector to a coaxial cable.

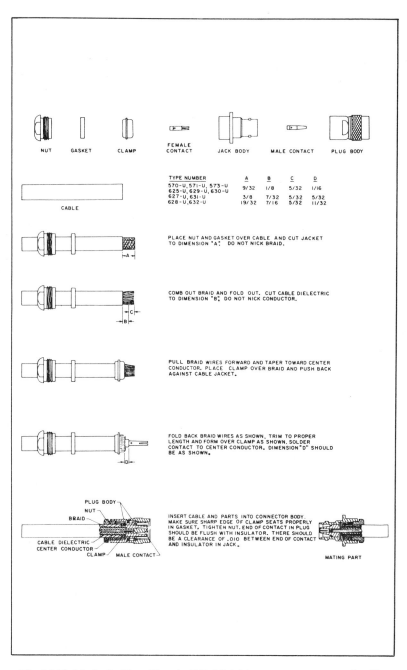

TYPE NUMBER	A	B	C	D
570-U, 571-U, 573-U 625-U, 629-U, 630-U	9/32	1/8	5/32	1/16
627-U, 631-U	3/8	7/32	5/32	5/32
628-U, 632-U	19/32	7/16	5/32	11/32

NUT GASKET CLAMP FEMALE CONTACT JACK BODY MALE CONTACT PLUG BODY

CABLE

PLACE NUT AND GASKET OVER CABLE AND CUT JACKET TO DIMENSION "A". DO NOT NICK BRAID.

COMB OUT BRAID AND FOLD OUT. CUT CABLE DIELECTRIC TO DIMENSION "B". DO NOT NICK CONDUCTOR.

PULL BRAID WIRES FORWARD AND TAPER TOWARD CENTER CONDUCTOR. PLACE CLAMP OVER BRAID AND PUSH BACK AGAINST CABLE JACKET.

FOLD BACK BRAID WIRES AS SHOWN, TRIM TO PROPER LENGTH AND FORM OVER CLAMP AS SHOWN. SOLDER CONTACT TO CENTER CONDUCTOR. DIMENSION "D" SHOULD BE AS SHOWN.

INSERT CABLE AND PARTS INTO CONNECTOR BODY. MAKE SURE SHARP EDGE OF CLAMP SEATS PROPERLY IN GASKET. TIGHTEN NUT. END OF CONTACT IN PLUG SHOULD BE FLUSH WITH INSULATOR. THERE SHOULD BE A CLEARANCE OF .010 BETWEEN END OF CONTACT AND INSULATOR IN JACK.

PLUG BODY
NUT
BRAID
CABLE DIELECTRIC
CENTER CONDUCTOR
CLAMP
MALE CONTACT
MATING PART

Fig. 16-20. Method of installing the UG-473A/U connector to a coaxial cable.

Cables used to conduct rf energy are of the coaxial type. These transmission lines can be made effective up to 3000 MHz. Coaxial cable, or "coax," is made up of a center conductor (sometimes hollow) and an outer conductor. A dielectric material separates the two. Coax for airborne installations employs one of three types of dielectrics. The air dielectric has insulated spacers to support the center conductor. The foamed-plastic type has excellent dielectric properties but lacks flexibility. A solid dielectric is used in most aircraft installations.

Coaxial-cable connectors must be designed to form a union with minimum attenuation and minimum disturbance of the standing-wave ratio. They also must be carefully assembled to the line. Figs. 16-19 and 16-20 illustrate the steps for assembling the UG-88/U and UG-473A/U connectors to a coaxial cable.

INTERFERENCE SUPPRESSION

Electronic systems must be guarded from electrical interference of any kind which will deteriorate usefulness or accuracy. Interference is usually in the form of random voltages or "noise." It originates either from sources within the airplane or from external phenomena.

The engine ignition system is the greatest source of noise within the airplane. The magnetos in particular are sources of powerful electromagnetic disturbances. To ensure prompt, hot ignition, the magneto pulses must have steep fronts. These transient currents cover a spectrum from 100 kHz to 300 MHz and hence can be a strong source of interference to electronic equipment on the airplane. Complete shielding is required to prevent radiation of these noise pulses, and only well-grounded shielding, which is free from resistance, will be effective.

There are two approaches to shielding. Either the interference is confined at its source, or it is excluded from places where it is harmful. Actually, both concepts are used. Obviously, electronic equipment should be housed in grounded, shielded cases.

Radial engines are shielded with a tubular manifold from which stem tubes or heavily braided shielded wire that connects to the spark plugs. The spark plugs, in turn, are enclosed in waterproof caps. The modern opposed-cylinder flat engines, developed for light aircraft, use integrally shielded plugs connected with braided shield harnesses. Magnetos and their associated assemblies are enclosed in metallic structures.

Ignition systems are not the only source of noise. Vibrators, commutators, relay switches, and other electrical devices can also be troublesome. In addition to proper shielding, bypass capacitors or filters in wires leading from engines or other equipment are sometimes necessary. An effective series filter at vhf can be obtained from a 5-microhenry choke shunted with a 0.005-microfarad capacitor.

Interference from external phenomena is more expensive and tedious to deal with. Static electric fields are formed on the outer aircraft surface. These fields discharge in complex ways and, in turn, induce random currents in electronic equipment. Surface fields are either *exogenous* or *autogenous*. Exogenous fields are caused by cloud activity, as in thunderstorms. Autogenous fields are caused by frictional contact of the aircraft surface with particles, such as snow or dust. The neutral particles deposit electrons on the aircraft surface, leaving the aircraft negatively charged. The amount of autogenous charge is a function of aircraft speed and frontal area and becomes present on both metallic and nonmetallic surfaces. Exogenous fields result because the aircraft becomes a conducting path between cloud fields.

Interference noise is generated when potential differences grow sufficiently to discharge. Discharge occurs either between surfaces or from surface to atmosphere. Three classifications of discharge are recognized:

1. Corona discharge.
2. Streamer discharge.
3. Sparkover.

Corona discharge occurs between sharp extremities such as wing tips, antenna protrusions, etc., and the atmosphere, as shown in Fig. 16-21A. "St. Elmo's fire" is a form of corona discharge, although corona discharge is not always visible. The frequency spectrum of this class of interference is shown in the aforementioned illustration. Antennas can be subjected to corona discharge. This can cause an impulse of high intensity to shock the first stages of a vhf receiver into brief transitory oscillation and, in turn, cause saturation. A short corona discharge pulse may thus appear as long as one millisecond. Successive corona discharge would then obliterate a signal.

Corona discharge may be reduced by lowering the resistance of the airplane to atmospheric fields. This is done with static dischargers or "wicks." These are carbon-impregnated cotton or nylon cords placed near the wing tips and other extremities. They should be installed as directed by the manufacturer.

Streamer discharge occurs on surfaces that are poor conductors, such as plexiglass windshields. Electrons build up on the surface, then discharge when the breakdown gradient of surrounding air is reached. Streamering appears between two charged regions on the surface or to a grounded part of the aircraft. Although the magnitude of streamering discharge is less than corona discharge, the spectrum is much broader as shown in Fig. 16-21B. Streamering may be reduced by using conductive paints on frontal surfaces, including radomes, and by employing especially conducting films on windscreen surfaces. On radomes, a compromise must be reached that will reduce streamering but not attenuate the radar signal beyond

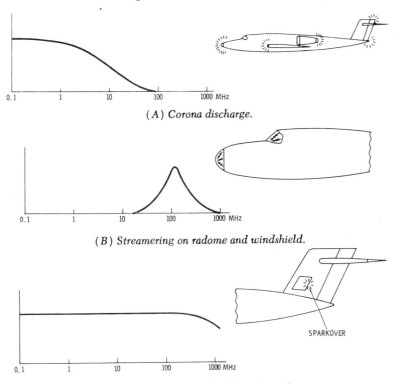

(A) Corona discharge.

(B) Streamering on radome and windshield.

(C) Sparkover on isolated metal panel.

Fig. 16-21. Illustration of three classes of static discharge with
accompanying frequency spectra.

a reasonable point. A resistivity of 1 to 100 megohms per square unit is recommended (since resistivity varies directly with path length and inversely with path width, any square unit of area holds a specific ohmic resistance).

Sparkover occurs between metallic sections that are electrically isolated, such as in access doors or even metallic paint and adjacent surfaces. The frequency spectrum of sparkover is related to the sparkover path. Sparkover may occur during periods when corona also exists. Characteristics of sparkover interference are shown in Fig. 16-21C.

Sparkover is reduced by careful bonding. Bonding not only reduces interference, but reduces the possibility of lightning damage by providing direct, low-resistance paths for currents, and also minimizes fire hazard. It usually consists of fine braids terminated in lugs bolted to the desired members. Copper bonding is used for interconnecting steel and aluminum alloy to aluminum. Direct contact between aluminum and steel is prevented by a cadmium coating that prevents corrosion. Bonding of engine, cowling, magnetos, fuel tanks, and occasionally even control wires, is essential. Most bonding is subjected to vibration or wear and tear, and thus should be periodically inspected. When surfaces are painted or repainted, they must be checked for electrical isolation and rebonded if necessary.

FLIGHT TESTING

Some form of flight test is required after a new installation, or after alterations and certain repairs. The success of the flight test will depend on the amount of preflight planning. Flight tests may be made by the pilot alone or, in large aircraft, by a complete crew including engineers and technicians. Nevertheless, whether the test is in a light airplane or a multiengine jet, some definite plan must be drawn up.

Where a flight test includes more than the electronics, still more coordination is necessary. Flight time is costly, regardless of the size of the airplane. Moreover, when schedules are not coordinated, the idle time of workers can increase labor costs unduly. For these reasons, a flight-test plan should be made. The first step is for each responsible technician to submit a schedule. The pilot or chief then draws up the test-flight plan in order to prevent conflicts. For example, the plan may call for high-altitude pressurization checks, but at high altitudes, obscured landmarks may make it impossible to check ADF. The duties of each individual during each portion of the flight are assigned. Alternate courses of action may be prescribed to be used in event of negative results on a certain test. The procedures set up in manufacturers' instruction manuals are considered standard; however, in most cases they are not necessarily mandatory.

SUMMARY

Airplanes flown for business purposes do not necessarily use the same equipment as airliners. Aircraft installations fall into three classifications. First, there are low-cost, panel-mounted systems that do not meet TSO. Second, there is a middle class that meets TSO. Third is the airline class of equipment. The term "one and one-half," often used in general aviation circles, applies to a system employing one

Fig. 16-22. The Genave GA-1000 is a "1½" NAV/COM unit that uses illuminated digital readouts for both navigation and communication. (*Courtesy General Aviation Electronics, Inc.*)

transmitter, a communications receiver, and another receiver for navigation (Fig. 16-22). The advantage of the "1 + ½" system is that the pilot may communicate with the ground without loss of VOR information. The time spent in changing a single receiver from communication to VOR station and vice versa is prohibitive. Hence the term "1½."

From this point, more sophisticated systems are built up. The use of a special central switching panel is desirable for switching over microphone and speakers, controlling volume, etc.

Heat is an important factor affecting the life of electronic equipment. Heat causes insulating materials to shrink and crack. Glass-to-metal seals become defective. Installations must be made to allow maximum heat dissipation. The effect of direct sunlight must be taken into consideration, even in flight. In larger installations, blowers must be utilized.

Index

A

Aerodynamic principles, 272-275
Aeronautical Radio, Inc. (ARINC), 323
Airborne weather radar, 227-247
  antenna stabilization, 237-238
  digital systems, 238-244
  ground mapping, 231-232
  performance analysis and maintenance, 244-247
  principles of, 228-231
  terrain collision avoidance, 231-232
  uhf techniques in, 234-236
Aircraft
  communication systems, 56-86
    cockpit audio requirements, 61-62
    maintaining and troubleshooting, 83-86
    microphones and headsets, 62-66
    receivers and controls, 66-76
      frequency synthesis, 68-70, 73-76
      selectivity, 60-61
      squelch operation, 67-68
    transceivers, 76-83
  installations, 344-377
    audio systems, 353-357
    cables and connectors, 368-373
    classes and categories of airplanes and equipment, 345-347
    electrical systems, 364-368
    equipment layouts, 348-353
    equipment standards, 347-348
    exhaust-gas temperature instrumentation, 357-358
    flight testing, 376
    fuel-monitoring instrumentation, 358-359
    interference suppression, 373-376
    standard sizes and form factors, 359-364
Airline Electronic Engineering Committee (AEEC), 323
Airport surface detection equipment (ASDE), 22-23
Air route surveillance radar (ARSR), 29
Air route traffic control center (ARTCC), 16
Air-surveillance radar (ASR), 20-22
Air traffic control (ATC), 14-15
Air Transport Radio (ATR) standard case sizes, 359, 362-364
Airways
  flying, 29-33
  structure of, 15-17
Altitude hold, 295
Altitude reporting, radar beacon transponders, 213-215
Angular momentum, gyroscopes, 250
AN radio range. See Low-frequency range system
Antennas
  goniometer systems, 170-172
  loop, 160-164
  requirements for aircraft, 52-55
  stabilization, 237-238
  wire, 52-53
Area navigation, 110
Automatic direction finders (ADF)
  basic system, 166-167
  circuitry, 167-170
  flight calibration, 175, 177-178
  frequency synthesis in ADF receivers, 172-173
  goniometer antenna systems, 170-172
  right/left direction finders, 164-166
  standard systems, 173-175
  testing and maintaining, 178-179, 336-343
Autopilots, 289-304
  for high-performance aircraft, 297-304
    approach mode, 303-304
    gyro mode, 302
    heading mode, 302-303
    NAV/LOC mode, 303
    yaw-axis control, 304
  simple stability systems, 289-292
  three-axis, 292-297

Radar—cont
  beacon transponders
    coding and decoding, 210-213
    Discrete Address Beacon System
      (DABS), 209
    false targets, 207-209
    fruiting interference, 208
    garbling, 207
    maintenance and testing, 222-226
    principles of, 205-207
    second-time-around targets, 208
    side-lobe suppression (SLS),
      209-210
  precision-approach (PAR), 18-20
  surveillance, principles of, 203-204
Radio compass. *See* Automatic direc-
      tion finders (ADF)
Radio spectrum, aviation, 41-55
  hf bands, 47-49
  low and medium frequencies, 45-47
  uhf bands, 51-52
  very-low frequencies, 44-45
  vhf bands, 49-51
Radio Technical Commission for
      Aeronautics (RTCA), 322
Radio waves, effect at aviation
      frequencies, 41, 44
Ramp testing, 329-330
Rate gyros, 250-251
Rate-of-turn indicators, 251-253
Readouts, solid-state, 317-320
Receivers and controls, aircraft, 66-76
  frequency synthesis, 68-70, 73-76
    heterodyne method, 68, 70, 73
    phase-locked-loop method,
      68-70, 73-76
  squelch operation, 67-68
Receiver, synchro, 255-257
Regulatory agencies, 321-323
  Aeronautical Radio, Inc. (ARINC),
      323
  Airline Electronic Engineering
      Committee (AEEC), 323
  European Airline Electronic
      Committee (EAEC), 323
  Federal Aviation Administration
      (FAA), 13, 322-323
  Federal Communications
      Commission (FCC), 323
  International Civil Aviation Organi-
      zation (ICAO), 322-323
  Radio Technical Commission for
      Aeronautics (RTCA), 322
Rho-theta navigation, 107-111
Right-hand rule for gyroscopes, 251

Right/left direction finders, 164-166
Roll/yaw stabilization, 293
Runway visual range (RVR), 24

S

Second-time-around targets, 208
Selectivity, aircraft receivers, 60-61
Semiconductor
  impurities, function of, 307-308
  junction, 306-309
Servo systems, 254-271
  command servo, 260-261
  error sensing, 259-261
  flux valve, 270-271
  loop(s)
    basic, 259-261
    theoretical analysis of, 267-269
  parameters, 261-266
    damping, 262-266
    hunting, 261-262
    stability, 261-266
    stabilization, 265-266
  servo computer, 261
  slaved gyro, 269-271
  synchro components, 255-259
    control transformer, 258-259
    differential, 258
    generator, 255-256
    receiver, 255-257
    transmitter, 255-257
  torque gradient, 258
Shop facilities and regulations,
      321-343
  certified repair station, 324-327
  FCC requirements, 323-324
  regulatory agencies, 321-323
    Aeronautical Radio, Inc.
      (ARINC), 323
    Airline Electronic Engineering
      Committee (AEEC), 323
    European Airline Electronic
      Committee (EAEC), 323
    Federal Aviation Administration
      (FAA), 13, 322-323
    Federal Communications Com-
      mission (FCC), 323
    International Civil Aviation Orga-
      nization (ICAO), 322-323
    Radio Technical Commission for
      Aeronautics (RTCA), 322
  technical manuals, 327-328
  test equipment, 328-343
    ADF, 336-343
    ramp, 329-330
    signal generators, 330-336

383